T0260179

The Abyss of Time

Ptolemaic orbits from *Harmonia Macrocosmica*, by Andreas Cellarius, 1661.

Other earth science titles from Dunedin
for the general reader include:

Volcanoes and the Making of Scotland (2015)
Second edition, Brian Upton, ISBN: 9781780460567

Edinburgh Rock: The Geology of Lothian (2006)
Paperback edition (2013) ISBN: 9781780460338

Death of an Ocean: A Geological Borders Ballad (2009)
Paperback edition (2013) ISBN: 9781780460345

For further details of these and other Dunedin
Earth and Environmental Sciences titles see
www.dunedinacademicpress.co.uk

The Abyss of Time

A study in geological time and Earth history

'The mind seemed to grow giddy by looking so far into the abyss of time.'
John Playfair (1748–1819) on seeing the unconformity at Siccar Point, read
1803, Transactions of the Royal Society of Edinburgh (1805)

Paul Lyle

Formerly lecturer in geology, University of Ulster

DUNEDIN
EDINBURGH ◆ LONDON

Published in the United Kingdom by
Dunedin Academic Press Ltd.
Head Office:
Hudson House, 8 Albany Street, Edinburgh, EH1 3QB
London Office:
352 Cromwell Tower, Barbican, London, EC2Y 8NB

www.dunedinacademicpress.co.uk

bitlit

A **free** eBook edition is available
with the purchase of this print book.

CLEARLY PRINT YOUR NAME ABOVE IN UPPER CASE
Instructions to claim your free eBook edition:
1. Download the BitLit app for Android or iOS
2. Write your name in **UPPER CASE** on the line
3. Use the BitLit app to submit a photo
4. Download your eBook to any device

ISBNs
9781780460390 (Hardback)
9781780465432 (ePub)
9781780465449 (Kindle)

© Paul Lyle 2016

*The right of Paul Lyle to be identified as the
author of this work has been asserted by him
in accordance with sections 77 and 78 of the
Copyright, Designs and Patents Act 1988*

British Library Cataloguing in Publication data
A catalogue record for this book is available from the British Library

Typeset by Makar Publishing Production, Edinburgh
Printed in Poland by Hussar Books

Contents

To Sylvia with love and gratitude – *bon ton roulet* – let the good times roll …

Acknowledgements

It is a pleasure to acknowledge those many people who have encouraged and helped me during the course of this project. Principal among those was John Arthurs; his comprehensive and perceptive comments on an early draft of the manuscript did much to improve the content and layout of the book and I thank him for his efforts. The comments from an anonymous reviewer were also extremely useful in the final revision of the text. I am grateful to Edward and Isa Ferguson for their helpful comments on parts of the manuscript and for again allowing me to use Anrhin as a writing retreat. My thanks go also to Rachel and Gary Smylie who provided invaluable photographic and computer support without which I would have been floundering.

My friends and colleagues among the diverse geological community in Ireland have always been free with opinions and comments, gratefully received in the spirit in which they were delivered, and I feel privileged to be part of such a vibrant network. My thanks to you all, you know who you are.

As always the contribution made by Sylvia Lyle has been immense and I thank her again for her constant support and encouragement. I appreciate her consistent willingness to pass comment on whatever piece of text I happen to be working on, even if I don't always subscribe to her analysis!

They also serve who cheer me on from over the water, so thanks to Gareth and Yvonne and Simon and Jen. Future time belongs to Beth and Jack.

Irrespective of all such support, any errors or omissions are solely my responsibility.

Preface

The idea for this book originated in a series of lectures for the general public which I gave in the Ulster Museum in Belfast in the autumn of 2005, on the theme of time and the geological history of Ireland. These lectures were well attended and the numbers of those turning up stayed high, even as the nights grew longer and the weather became more inclement. This suggested there was an interest in the concept of time in geology that was not being met elsewhere at that time. Hutton's famous phrase – 'we see no vestige of a beginning, no prospect of an end', although flawed, clearly struck a chord with many of the audience. Perhaps for the first time they were thinking directly about the extent of time in Earth history, of time passing and of their place in the continuum of time, if indeed it is a continuum. Geologists, by the nature of their training, perceive time in a different way from non-geologists. A view of any landscape is automatically interpreted in terms of the age of the rocks that form that scenery and the landscape is also interpreted as a sequence of events in Earth history. This ability to 'read the landscape' adds a further dimension to the appreciation of the world around us but also allows a greater understanding of our place in the grand scheme of things, on Earth, in the Solar System and in the universe. It is clear that there is a poor understanding in general of the extent of geological time.

Recent surveys taken in the USA have shown that almost half the population there believe that human beings were created in their present form around 10,000 years ago. There is also a poor understanding of the order in which events in Earth history took place. Many people believed that certain events had occurred simultaneously when in fact they were separated by a considerable amount of time. A good example of this is the widespread belief that dinosaurs and humans were on the Earth simultaneously. A further cause for surprise is that, as well as this level of ignorance about time, there was also no apparent concern over their lack of knowledge among those polled. It seems that those enthusiasts who turned out for my lectures were very much an interested minority.

It is not just for aesthetic reasons that people should appreciate geological time, its extent and its ramifications. The world we find ourselves living in today is one of expanding populations using increasing amounts of resources in the form of fuels, often fossil derived and carbon based, to the extent that if present trends continue there will be insufficient resources to sustain the Earth's population. The importance of understanding geological time is that many of the world's resources result from cyclical processes. These cycles are sometimes relatively short term, for example the water cycle whereby water evaporates from the oceans to form clouds that come over land and produce rainfall which makes its way back to the ocean via streams and rivers and so completes the cycle. This occurs over periods as little as a few months in some places but over hundreds or thousands of years for some of the large-scale underground water sources such as the Ogallala Aquifer in the mid-western states of the USA. This aquifer, like many such reserves worldwide, is being over-used. The rate of discharge from the aquifer is greater than the rate of recharge, and so the cycle is not being completed. Water can probably be considered as a 'renewable resource', but that term cannot be applied to resources such as copper or oil which result from geological cycles which take hundreds or thousands of millions of years to complete. The oil and gas being so greedily consumed in the world's vehicles and central heating boilers mostly began to form over 100 million years ago and when the reserves are used up we will have to wait a very long time for their replacements to be available.

A greater appreciation of these timescales by politicians and their constituents might help them to agree measures to ameliorate the catastrophe we seem to be heading for. To understand whether or not we are heading into a period of human-induced climate change, with perhaps disastrous effects for the whole of Earth's population, it is essential that we examine major climate changes that have occurred throughout geological time. This requires an appreciation of geological history, with time measured not just in a relative way – this rock is older or younger than that rock – but also a time framework that tells us the age of a bed of rock, or a mud layer on a lake bed, or a layer of ice in the Greenland ice cap, in as precise a number of years as possible.

So, who is this book aimed at? One obvious interest group is the cohort of people such as those who came to my lectures. They were people who are interested in their surroundings and their environment, who are seeking the intellectual challenge presented by confronting the extent of geological time and the implications that may have for their view of themselves and their place in the universe. If I am feeling optimistic I think that perhaps these people are not such a small minority after all. There is clearly great interest in matters geological, archaeological and astronomical, to judge from the diet of popular television programmes available, and the numbers of books on popular aspects of science that are published and purchased every year. If it is possible to popularize the concept of 'black holes' in astronomy, then the idea of an Earth that formed 4.5 billion years ago is not so daunting.

However, engaging with that particular constituency could be regarded as preaching to the converted. The challenge is to engage the majority of the population who up to now have shown little interest in the topic. I believe this group is as big as it is because geologists as a profession have not been as good at communicating with non-geologists (the rest of the world!) as they have been at talking to other geologists. They have been guilty of a lack of outreach and if people do not fully appreciate the role of geologists in society, then geologists themselves are partly to blame for keeping their professional secrets too much to themselves. So this book is an attempt at that outreach to those members of the population who have been previously neglected by our profession, people who are concerned about many of the environmental problems we are facing today and in the near future, but who do not as yet have all the information they need to make an informed judgement on the best way forward. It is also aimed at decision-makers, those members of the legislative and executive branches of governments who are required to make and administer the laws. Such legislation is quite often drawn up with a shocking level of ignorance about the longer term implications of what is being proposed. The book is aimed at raw material producers and those in the manufacturing industries, the suppliers and users of precious resources. An appreciation of the workings of geological time and processes must surely be helpful when balancing the needs of consumption and conservation.

There are many academic and professional fields nowadays that have a peripheral engagement with geology – environmental scientists, geophysicists, water scientists and civil engineers for example – for whom a grasp of the role of the fourth dimension would be useful. For students or practitioners in these areas this book deals with the principles of relative and absolute time, treatment of which is often absent from conventional university courses.

And then there are those people who know little or nothing about geological time, who have never thought about it and furthermore see no reason why they should. Some of these people resist the idea of a very old Earth because of their religious beliefs, a view that has been maintained since the early days of the Scientific Revolution when the ideas of scientists such as Galileo were seen as a challenge to the Church authorities of the time. My own experience suggests that faith invariably trumps scientific reason so I would not anticipate any major breakthrough in that group by this book alone. An acceptance of the extent and ramifications of geological time may only come when the implications of a shortage of non-renewable resources become obvious.

Perhaps the principal aim of this book is to enable non-geologists to share in what geologists have tended to keep for themselves too long. To share that appreciation of what geological time means in enabling us to put ourselves in context within the historical events of the Earth and its birth from the solar nebula about 4,500 million years ago and to be able to look at a landscape and interpret the sequence of events that produced it.

Dr Paul Lyle
C. Geol, FGS
December 2014

Chapter 1

In good time

If you free yourself from the conventional reaction to a quantity like a million years, you free yourself a bit from the boundaries of human time. And then in a way you do not live at all, but in another way you live forever...
John McPhee, *Basin and Range*, 1981

In good time, time is of the essence, about time... how many commonly used expressions mention the passage of time and the measurement of time? Modern life for many people revolves around a tight schedule of events governed by a seemingly implacable clock that will accept no excuses. And yet time can often appear to be less than absolute, those occasions when time seems to pass very quickly, when our senses are heightened by pleasure or fear, other occasions when the clock is being watched and time passes excruciatingly slowly. The passage of time is surely not perceived to be the same by the child at the end of the school summer holidays, puzzled at where all that wonderful time went to so suddenly, and the long-term prisoner contemplating the seemingly never-approaching end of sentence.

Geologists, along with cosmologists and astronomers, are scientists who must factor in the added dimension of *time* to their investigations and this book aims to investigate geological time, but does geological time differ from other sorts of time? We categorize time in our daily lives by the use of such terms as working time, leisure time or down time. Should these different categories be measured and evaluated in the same way, do they pass at the same rate? Without straying too far into the realms of physics it is clear from the work of Einstein in the first half of the 20th century that time and space are connected in a way that is perhaps not immediately apparent in our everyday lives – the idea that time flows more slowly for objects moving at close to the speed of light is a puzzle to most people. Many of the processes investigated by geologists took place millions and sometimes billions of years ago, and may have lasted over periods covering many millions of years.

In the early history of the Earth there is clear evidence that the length of day on Earth was not the same as today – the length of the day on the early Earth around 3.5 billion years ago (3.5 bya) was about 9 hours and there were over 1,000 days in the year. The rotation of the Earth has been slowing gradually with time due to gravitational effects caused primarily by the arrival of water on the Earth's surface from about 3.5 bya. How do these changes affect our measurements of time passing?

Since our species *Homo sapiens* first evolved there has been a fascination with time, its measurement and its significance. Before attempting to understand the various ways time has been viewed by society through the ages we need to consider the importance of time to the science of geology. Geologists look at the landscape in a way that differs from the general population. John McPhee, quoted at the head of this chapter, although not a geologist himself, recognized this when he said 'Rocks are records of events that took place at the time they formed. They are books. They have a different vocabulary, a different alphabet, but you learn how to read them'.

The Geological Society of London is the learned society and professional body for the Earth sciences in the UK and is the oldest such society in the world, founded in 1807. With an apt motto – '*Quincquid sub terra est*' or 'Whatever is under the earth' – it is well placed to define the functions of geologists. It argues that an understanding of the interaction between environmental change and the evolution of life over hundreds of millions of years gives geologists a valuable perspective on the changes that humans are now causing by burning fossil fuels and on our wider impacts on the planet. Geologists were primarily involved in extracting resources from under the ground and somewhat ironically will in future play a significant role in reducing the impact of carbon emissions by putting carbon dioxide and radioactive waste back underground where they originally came from.

The Irish poet Seamus Heaney rather more elegantly encapsulated the essence of geology when he talked about its 'vastness and pastness' and described his view of the role of a geologist in the following perceptive terms:

> Geologists face the facts before they force the issues. Their first concern is with the here and nowness of the world or its there and then-ness, although this may well lead to questions of its how and whyness. They read the ground rules of earth itself and tell the time by the planet's body clock. (Conference proceedings, Dublin Castle, 2002.)

A geologist interprets the landscape by recognizing the rock types present, the likely sequence of their formation and their spatial distribution over the area. Some landscapes are easier to read than others. Monument Valley in Utah

Figure 1.1 Rock formations, Monument Valley, USA. © Andrew Zarivny/ Shutterstock

in the USA, the scene of many epic western movies, is a landscape largely unaffected by the camouflaging effects of vegetation and shows clearly its rock components and structures (Fig. 1.1).

Individual rock types and their relationships to each other can be read like the pages of a book; all that is required is to learn the language the book is written in and the narrative will unfold.

A highland story

Figure 1.2 shows three prominent peaks on the skyline in the Assynt district of the north-west highlands of Scotland. The scenery is unique in Britain, and consists of a series of spectacularly isolated mountains rising abruptly above an undulating heather-covered plain dotted with innumerable small lakes or lochs.

The spectacular scenery shown in Figure 1.2 covers much of the early geological history of Britain and can serve as an illustration of the concept of deep time in geology, providing we take the advice from John McPhee at the head of the chapter and free ourselves from the conventional reaction to a quantity like a million years. This landscape tells a story that encompasses many hundreds of millions of years and its understanding and appreciation may take some mental readjustment for readers not accustomed to the sorts of numbers bandied about by geologists. This scenery is the product of geological forces acting over the last 3,000 million years (3,000 myr) or so and has a great deal to tell us of what took place on this particular part of the Earth's crust during that period.

Figure 1.2 The Torridonian sandstone peaks of Canisp, Suilven and Cul Mor, Assynt district, north-west Scotland.

The isolated peaks such as Suilven (Fig. 1.3), are termed inselbergs, literally 'island mountains', for their resemblance to islands in a rock sea. In this part of Scotland the mountains consist for the most part of two distinct layers. On top is a series of reddish-purple sandstones, known to geologists as Torridonian and forming the steep-sided part of the mountain. This sits on a very different rock, called gneiss, designated as Lewisian after the Isle of Lewis in the Outer Hebrides and forming the undulating landscape seen in the foreground. The gneiss forms a peneplain, meaning a surface which has been worn down nearly flat by erosion. Gneiss is a metamorphic rock, one which has been changed by being subjected to high pressures and temperatures deep in the crust, showing clear signs of mineral banding and folding (Fig. 1.4A). The red Torridonian sandstones occur

Figure 1.3 The Torridonian sandstone peak Suilven with the Lewisian gneiss peneplain in the foreground.

Figure 1.4 **A**: Folding and banding in Lewisian gneiss. **B**: Layers of Torridonian sandstone.

as a series of separate layers (Fig. 1.4B), each up to a few metres or more in thickness, usually tilted a few degrees below the horizontal, but unlike the gneiss they show no signs of folding or distortion.

To grasp the notion of deep time let us consider an observer who has made the trek across the rocky sea of the Lewisian peneplain in the foreground of Figure 1.3 and has managed the steep climb up to the mountain's summit.

5

People are familiar with the idea that the stars are at very great distances from us, and that these distances are so vast that they can only be measured in light years. A light year is the distance that light can travel in a year at 299,792 kilometres (km) per second, a distance of around 9.5 trillion kilometres. If an object is 1,000 light years away then the light seen from that object is 1,000 years (1 kyr) old. Our Sun is about 8 light minutes away – if it was suddenly to explode we would not know about it until 8 minutes after the event. In this way there is a relationship between distance and time, and our intrepid climber on the summit of Suilven, gazing up at the night sky is therefore looking back in time. In a similar way, however, peering down from the summit is also looking back in time. The sequence of rocks can be read as a series of events or a time-line. In this case there is a connection between time and the three-dimensional relationship of the rocks. Someone looking down from the red Torridonian sandstones on the summit of Suilven to the Lewisian gneiss at the base of the mountain is also looking back in time, to processes occurring between approximately 900 million years ago (mya) to more than 2,500 mya.

The history of this part of Britain over that period involves the building of mountain ranges and their destruction, climate changes ranging from hot deserts to frozen tundra, and a journey across the Earth's surface from the southern hemisphere near the South Pole to its present position around latitude 58 degrees North. These events and the order in which they happened make up the geological history of an area and it is the job of the geologist to work out that history as fully as possible. The story of this landscape has been unravelled by successive generations of geologists since the middle of the 19th century, and by combining the evidence gathered in the field with data collected in the laboratory it is possible to read that sequence of events in some detail.

Consider the Earth sometime between 2,500 and 3,000 mya. In appearance the arrangement of continents and oceans was very different from the view of our planet that we have become accustomed to in the last 50 years or so as photography from space became commonplace. The distribution of continents and oceans on the surface of the Earth is not fixed. Instead, due to a process called plate tectonics, ocean basins open and close over periods of hundreds of millions of years. Continents split apart and move over the surface of the globe before colliding with other continents to form new continental areas. Around the time of formation of the oldest rocks in our photograph, the Lewisian gneisses, the main continents had amalgamated into a supercontinent containing rocks from the present continental areas of north-eastern Canada, Greenland, Scandinavia and the Baltic countries. At this stage in Earth's history there was no plant or animal life on land and life in the oceans consists of relatively primitive organisms.

About 1,000 mya the Lewisian gneisses formed a landscape very similar to the one they form today, a low-lying undulating surface resulting from a prolonged period of erosion. This surface was gradually buried under a great depth of sand, brought in by a major slow-moving river system. The story, however, does not start at 1,000 mya because the gneisses formed around 2,500 mya. We know this 2.5 billion year (byr) age of the gneisses by measuring ratios of certain radioactive elements present in the rocks and they are some of the oldest rocks in Britain or Ireland. The gneisses are metamorphic: they were another rock type before being changed by heat and pressure. In this case it seems likely they were granites and sandstones. Since the age calculation tells us only the age of the metamorphic changes, then these original rocks must have been more than 2,500 myr old when they were formed. We also know from their mineral compositions that the rocks were buried to a depth of perhaps 30 km or more, probably during an early period of mountain-building. From 2,500 to 1,000 mya erosion gradually brought these deeply buried rocks to the surface. Then, gradually, this ancient gneiss landscape was buried under the Torridonian sandstones so that by 800 mya there was a maximum thickness of around 5,000 metres (m) of sandstone which completely covered the Lewisian gneiss. The early part of the Torridonian was deposited in a hot arid climate as indicated by the mainly red colour of the sandstones, but by the time the younger beds were being deposited the environment had become wetter. The Torridonian sandstones were then in turn covered by younger rocks, pale sandstones that can be seen at the summits of other peaks in the area such as on the peak Quinag (Fig. 1.5).

The succession or order of rocks in this part of Scotland, from oldest at the bottom to youngest at the top, is:

- Cambrian sandstones (grey coloured) about 500 myr old youngest;
- Torridonian sandstones (red coloured) about 800 myr old;
- Lewisian metamorphic rocks (banded gneiss) about 1,500 myr old oldest.

These Cambrian sandstones were deposited as the continent was gradually flooded by a new ocean which was forming and expanding. The geological processes associated with this ocean, called Iapetus, were to have major effects on the geology of Britain and Ireland in succeeding years and have moulded the structure of the British Isles that we are familiar with today. Following this maximum of deposition the sandstone beds began to be eroded by a fast-flowing river system, which gradually wore them away over hundreds of millions of years.

The last few tens of millions of years have seen a continuation of erosive conditions in northern Britain, with faster flowing river systems and a

Figure 1.5 Quinag showing Torridonian sandstone over Lewisian gneiss, with a cap of younger, grey sandstone. © Targn Pleiades/ Shutterstock

prolonged ice age. This most recent glaciation in the last million years or so was responsible for major landscape changes over most of the British Isles. The cumulative result of these erosive forces in northern Scotland has been to strip off the great thickness of Torridonian sandstones. The ancient Lewisian surface has thus reappeared after being buried for 800 myr, leaving the inselbergs like Suilven as remnants of a much more extensive covering of Torridonian sandstones.

Thus the landscape can reveal its secret history if we know what to look for. In this case it also illustrates two contrasting aspects of time. First is time moving in a straight line, a sequence of unique events with each happening being followed by the next throughout geological history. The formation of the supercontinent was followed by its break-up and the subsequent formation of a new ocean. Time, however, also moves in cycles. In this case the Lewisian rocks had been eroded to a peneplain surface 1,000 mya at the end of a cycle of erosion. This surface was then buried by the Torridonian sandstones as part of a cycle of deposition that followed the erosion cycle. That billion year old surface has been rediscovered as the erosion processes that perpetually renew the Earth's surface removed the 5,000-m-thick blanket of Torridonian sandstones. Imagine the Lewisian gneisses emerging, blinking, into the Scottish sunlight as they are gradually uncovered after about 800 myr in darkness, while the debris thus removed is transported by today's rivers and dumped in the sea to set up the next stage in the cycle of time.

Chapter 2

Tempus fugit – time flies

The Sun is new each day...
Heraclitus of Ephesus (544–483 BCE)

The science of geology is distinctly associated with the concept of geological time and its measurement, but as already noted time can be perceived and measured in very many different ways. The stone circle at Callanish on the island of Lewis (Fig. 2.1) is made up of huge blocks of gneiss similar to that seen at Suilven and discussed in the previous chapter. Monuments such as these were erected in many places in western Europe during the Neolithic period about 8,000 years ago (8 kya) and are part of an increasingly complex relationship with time that developed as society moved into the Bronze and Iron Ages.

Figure 2.1 Callanish stone circle, Isle of Lewis.

Time has a different meaning to those of us living in the developed world in the 21st century compared to the Neolithic populations who built the stone circles on Lewis thousands of years ago. To them time was measured by the daily, monthly and yearly cycles of the Sun and Moon. The world in which we live appears at times to be timetabled to the minute, and yet clocks only became important around the end of the 14th century in Europe, and these early versions possessed only an hour hand. Phone calls now are commonly charged by the minute, so at what stage in society's development did it become necessary to be able to measure a period of time as short as 60 seconds?

For as long as mankind has been aware of the passage of time its extent has been the subject of intense and often acrimonious debate. In the literary genre of science fiction, the concept of 'deep space' is long established and refers to that region of the universe outside not only our Solar System but outside our Galaxy. Conventionally in science fiction these areas are only accessible by using spacecraft which possess devices such as hyperdrives and which have the ability to travel at the speed of light. Since this facility is still firmly in the realms of fiction we do not need to consider the concept any further, but need rather to expand the 'deep' concept to the subject of this chapter – time. The idea of deep time is a relatively new one and encapsulates the idea that geological time is vast and measured in millions, if not billions, of years. The term was applied to geological time and geological history by John McPhee in his general interest study of the geology of North America – *Basin and Range* – published in the early 1980s. This concept of deep time is in sharp contrast to the previously widely held view that the age of the Earth could be measured in a few thousands of years. A biblical timescale for the origin of the Earth and a geocentric view of the universe were important tenets of faith in the Middle Ages, contradiction of which was dangerous for any philosopher of the period. The idea of the Earth in orbit around the Sun rather than vice versa and the concept of a nearly limitless age for the Earth were twin pillars of what has been described as the Scientific Revolution, beginning in the mid-16th century and laying the foundations of modern science. As a way of illustrating the immensity of geological time and the relatively late arrival on the scene of humans, McPhee imaginatively likened the whole of geological time to the length of an outstretched arm. On that scale the whole of human history could be removed by a single stroke of a nail file across the middle finger. The Boulder Community Network in Colorado offers a more numerical illustration of deep time (see Fig. 2.2).

The age of the Earth is 4,500 myr. How we come to this figure will be dealt with in later chapters. If the rectangle at the top left of the diagram above was 100 m on its long side and this represents 4,500 myr of Earth history, then 1 millimetre (mm) would equal 45 kyr. This last millimetre can be subdivided into

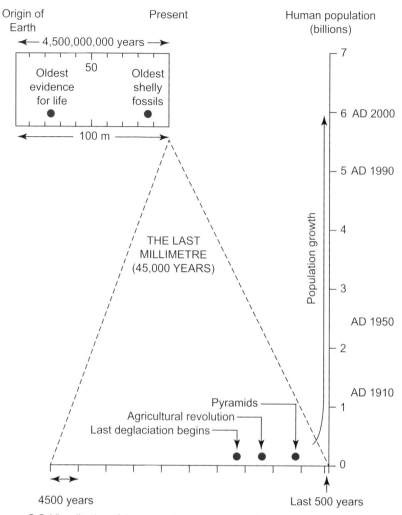

Figure 2.2 Visualization of deep time, from the Boulder Community Network, Visualizing Deep Time. The age of the Earth, 4,500 myr is represented by 100m. The last 1mm of this equals 45,000 years.

ten units of 4.5 kyr each (see the horizontal line at the base of the diagram) and used to mark major events during that period. This shows the last de-glaciation beginning about 20 kya, the beginning of the Agricultural Revolution around 12 kya and the date for the building of the pyramids at around 5 kya. The diagram clearly illustrates that the influence of humans on the Earth only started with the advent of agriculture, in roughly the last quarter of the last millimetre of the 100 m representing geological time in its entirety. While this can be a sobering thought as to the significance of our contribution to date, the rate at which the human population is expanding (shown on the right of the diagram) has serious implications for our future on the planet.

Various ways have been tried to illustrate the 4.5 byr of Earth history in a way that is meaningful to the lay-person. Devices such as using 24 hours to represent the fullness of geological time mean that the first living cells did not appear until around 7.00 hours. The 'explosion of life' when a wide variety of complex life forms appeared in a relatively short time around 500 mya happened at 21.20 hours on our clock and the development of hominid forms, leading eventually to us, *Homo sapiens*, began about one minute to midnight (see Fig. 2.3). If nothing else this format illustrates clearly the very short period of Earth time for which we have been around. The paltry few million years of hominids is over-shadowed by the hundreds of millions of years that species such as the sharks have been swimming in Earth's seas. Another commonly used device to depict geological time is the spiral of time where time is measured outwards from a central origin in a spiral form (Fig. 2.4).

Imaginative techniques such as these are useful in that they allow the non-geologist ways of reducing mostly incomprehensibly large numbers of years such as 4.5 billion, or 4,500,000,000, or 4.5×10^9 (all ways of expressing the same number), to something they can relate to.

A further source of confusion when trying to explain the concept of geological time to non-geologists is the difference in the orientation of timelines between historical events and geological events. The timeline in Figure 2.5 shows the principal landmarks in human development over the last 50 myr. The line is horizontal with time flowing from left to right. In geology, however, timelines are always vertical in orientation and the most recent event in a geological timeline is always at the top as we discussed when dealing with the rock sequence in the north-west highlands of Scotland (see Fig. 1.5). The confusion engendered by this difference in timeline orientation between historical events and geological events is compounded by the idea that geologists perceive time as flowing backwards to the past along their timelines, while for the rest of humanity time in their daily lives is flowing forwards to the future.

Speculation on the extent of time is not a new phenomenon but has exercised enquiring minds for millennia. As in many other fields such as astronomy and mathematics the musings of the early Greek philosophers on time and its passing were to have an influence that lasted until the Middle Ages and often beyond.

The Greek historian Herodotus, considered by many as the founding father of history, was aware in the 5th century BCE (Before Common Era*) that river deltas such as the Nile in Egypt (Fig. 2.6) were formed by the accumulation of mud and silt particles deposited by the river each year when it flooded.

*Common Era or CE and Before Common Era or BCE are alternative names of the calendar eras Anno Domini (AD) and Before Christ (BC) respectively. These terms are increasingly used in archaeology.

© Artur Synenko/ Shutterstock

Figure 2.3 Digital clock showing the approximate time of arrival of Homo sapiens on Earth, if the full extent of geological time is represented by a single 24-hour day.

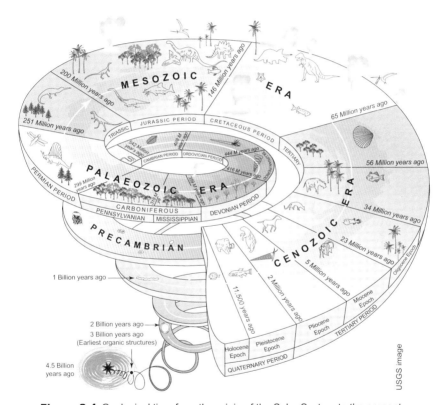

USGS image

Figure 2.4 Geological time from the origin of the Solar System to the present.

Recognizing the large scale of the feature he estimated that it had taken the Nile tens of thousands of years to build, a formidable span of time to the ancients, but more importantly a major intellectual leap forward when one considers that nearly 2,500 years later there are still many people in the world who would not accept that sort of age for the entire Earth, much less a single delta-forming event.

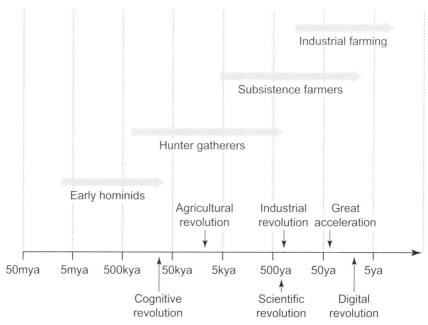

Figure 2.5 Simplified timeline showing principal landmarks in human development over the last 50 million years; mya: millions of years ago; kya thousands of years ago; ya years ago.

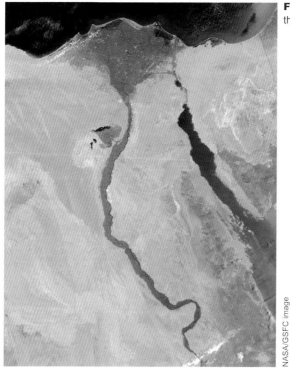

Figure 2.6 Aerial view of the Nile Delta, Egypt.

In contrast to this approach the philosopher Heraclitus of Ephesus, born around 535 BCE, said profoundly 'one cannot step into the same river twice' and considered the world to be composed of ceaseless movement and transformation. Around 100 years after Herodotus, Aristotle, thought by many to be the greatest of the early Greek philosophers, was putting forward the concept of a changeless or cycling eternity or infinity. The contrast here between the observational and empirical approach by Herodotus and the philosophical treatment by Heraclitus and Aristotle has been mirrored in the various controversies concerning time and its measurement since then. Aristotle's view of the cosmos (the word in Classical Greek meant world or universe, but also order as opposed to chaos) based on the Greek ideal of the circle as perfection was to influence astronomical thought until the time of Galileo in the mid-17th century. By this time medieval philosophers and theologians had developed the idea of the universe having a finite past with a starting point. This was based on a creation myth, something that is shared by the three main monotheistic religions, Judaism, Christianity and Islam. These contrasting philosophical approaches meant that from the very earliest times there was a strong contrast in the way that time was viewed. Either existence was timeless or eternal, or time was about the reality of change with the world continually in flux and consisting of ceaseless movement and transformation.

The Cognitive Revolution

This fascination with the passage of time, however, did not start with the Greeks and there is evidence that the earliest populations of *Homo sapiens* were aware of the various cycles of time in their environment.

Around 2.5 mya the genus *Homo* had evolved and by 2.0 mya various human species had spread from Africa to Eurasia. The Neanderthal species (*Homo neanderthalensis*, named after the Neander Valley in Germany) evolved from around 500 kya, preceding the origin in East Africa of our own species, *Homo sapiens*, by some 300 kyr. For more than 100 kyr *Homo sapiens* existed as a hunter-gatherer on the plains of East Africa, one of several variations on the human species and not apparently significantly different from any of the other higher primates. However, sometime around 70 kya our species underwent the first of a series of revolutions that would lead eventually to our present status as the dominant life form on the planet.

This change has become known as the Cognitive Revolution and refers to the ability to show creativity and imagination and to begin to communicate in an unprecedented way. The results of these changes in the thought processes of *Homo sapiens* include increasingly complex tools, sharing or exchanges with other groups, figurative art, music and dance, cooking and burial rituals.

The most commonly believed theory for this revolution in thought processes suggests that accidental genetic mutations altered the wiring of the brains of *Homo sapiens* enabling us to think differently, leading eventually to the emergence of modern languages. These superior communication skills meant *Homo sapiens* was capable of greater cooperation with larger numbers of individuals and thus able to plan and execute complex actions such as hunting. In addition the development of fictive languages (i.e. languages capable of expressing imagination and creativity) meant that as a species we were able to begin to build cultures. Among other innovations, myths, legends and religions all arose after the Cognitive Revolution.

This period in Earth history, known to geologists and archaeologists as the Palaeolithic, saw developments we would recognize now as the beginnings of civilization. These included the use of burial rituals for the dead, early examples of art and music, the wearing of clothes and, importantly for our discussion here, the first attempts to mark the passage of time.

The first myths and legends after the Cognitive Revolution would undoubtedly have included observations based on the yearly and monthly cycles of the Sun and the Moon. As long as 30 kya there are engravings on bone fragments from the Dordogne in France which have been interpreted by some archaeologists as an early record of the cycles of the moon and possibly the first primitive calendars (Marshack, 1972 – see Fig. 2.7). However there is no consensus on this among archaeologists. Other workers regard this and similar examples as tallies of events or pieces of moveable art, rather than a device for making predictions (Pásztor, 2011). However that does not mean that the Palaeolithic people were not interested in the sky or in marking the passage of time. Elsewhere around the same time sequential notches in sticks and periodic engravings on flat stones are testimony to an increased temporal awareness. Awareness of time passing and the realization and anticipation of the cycles of nature were seminal landmarks in human evolution. These cycles vary in scale from the daily rising and setting of the Sun, to the lunar cycle clearly shown in the night sky by changes in both

© Emilia Pasztor

Figure 2.7 Engravings on bone fragments from Abri Blanchard in the Dordogne, France.

shape and position of the Moon. Superimposed on this is the longer yearly cycle of the Sun represented by the change of the seasons. Observation of these events led to early societies absorbing the concept of time and incorporating it into society and culture. In the great cities of the Mesopotamia region of the Middle East 4–5 kya the first true astronomers came from the priestly caste who studied the movement of the heavenly bodies for astrological and religious reasons. In order to do this long-term records of celestial events were required. For example to track the orbit of Saturn requires at least a generation's worth of continuous observation. Cosmic time and human time were thus explicitly joined as astronomer priests began to keep track of long-term celestial motions across many human lifetimes.

The most fundamental astronomical cycle we experience is the rotation of the Earth about its own axis to give the variation from day to night and back again – governing our circadian rhythms – those rhythms repeated every 24 hours.

The cycle defined by the Moon (see Fig. 2.8) shows two distinct aspects – the variation of its position in the sky and the changes in the shape of the Moon we recognize as the phases of the Moon. It waxes from a crescent new moon to full moon and then wanes to gibbous and finally crescent again before the cycle restarts. The lunar cycle, which defines roughly a month, is a convenient measure of celestial time – celestial defined as pertaining to the heavens – and next to the daily cycle the one that impinges most readily on human consciousness. It is short enough to be easily counted but long enough to measure durations of many weeks. The expression 'many moons' as a measure of a long time may be a cliché familiar from Hollywood westerns, but nonetheless many early cultures kept lunar calendars rather than solar calendars.

The yearly cycle defined by the orbit of the Earth around the Sun which gives us seasonal change is the longest repeating period imposed by the heavens that

© Fluidworkshop/ Shutterstock

Figure 2.8 Moon phase cycle, from 1 day to 28 days old.

most people are aware of. Even in modern developed societies that are in so many ways removed from natural phenomena most people are aware of the changes inherent in the yearly cycle of the Sun. They know that it rises in the east, sets in the west, is higher and warmer and with longer days in the summer, lower and weakest in winter when the days are shorter. All of these represent the Sun's movement across the sky during the course of a year. The occurrence in many cultures of major festivals at the winter and summer solstices recognizes two of the main points in this cycle.

Observation of the various cycles around them allowed the early societies to develop the concept of time so that it could be said that time arose from the growth of culture.

The change from the earliest societies of Palaeolithic (Old Stone Age) times to the Neolithic (New Stone Age) centred on a warmer world, as the effects of the Ice Age receded, and the practice of farming spread, with its attendant changes involving permanent housing and the domestication of animals. This change in society from a largely hunter-gatherer mode of life to the more sedentary life of agriculture marked a radical change in society's engagement with its environment and to a change in how time was viewed and measured.

This change was marked by societal/cultural innovations including:

- food production through domesticated plants such as wheat, barley, flax;
- use of tools such as axes and grindstones to clear forest and plant and harvest crops;
- domestication of animals such as sheep, cattle, pigs;
- emergence of settled village life with permanent dwellings;
- the interment of the dead in cemeteries, sometimes featuring monumental tombs;
- the development of long-distance procurement systems for raw materials.

Thus a fundamentally new way of organizing human activities was required, as well as a new way of imagining culture and its place in the cosmos. Also each innovation required a daily engagement with and recognition of time unlike anything that had come before.

Material engagement

For the people in the Neolithic era, as in the earlier Palaeolithic and Mesolithic, the increasing use of the resources available to them from their surroundings would lead to a new understanding of time – for example the mastering of techniques in pottery-making and metallurgy and the technologies involved in planting and harvesting as agriculture developed. The spread of farming westwards and northwards from the Middle East led to the recognition of the daily and seasonal cycles of agriculture. The Neolithic was marked by major changes in the landscape brought about by this development of an agrarian lifestyle. Roughly in parallel with the growth of agriculture was the appearance of monuments in the landscape, the megaliths, including stone circles, standing stones, cairns and burial chambers. These monuments were probably built for a variety of purposes including memorials, territorial markers and as part of religious ceremony, but they represent a profound change in how the Neolithic people viewed time. Construction of the great monuments, which clearly involved huge commitment of resources of time and effort from the societies which built them, indicates an awareness of time past, with the presumed memorial function of the megalith.

Monument construction also, crucially, shows an awareness of time in the future, since the monuments were clearly built to long outlast the builders. The monuments at Stonehenge (Fig. 2.9) and Newgrange (Fig. 2.10), as well as

Figure 2.9 Stonehenge, England. © Walencienne/ Shutterstock

Figure 2.10 Newgrange, Ireland. © Pecold/ Shutterstock

acting as burial sites, were used to celebrate astronomical phenomena such as the summer and winter solstices, part of the yearly cycle of the Sun. The change in the Neolithic from a hunter-gatherer lifestyle to that of a farmer meant that there was a switch to a way of life marked by daily cycles and seasonal cycles and this is reflected in how they used constructions like Newgrange or Stonehenge to mark the highlights of the Sun's year. The orientation of the tombs at Newgrange (and also presumably at similar sites such as Stonehenge) may be linked to recording the critical turning points of the year – the winter and summer solstices and the equinoxes. This indicates a concern with control over the knowledge and ordering of time which was linked to the ancestral world by being 'captured' in the tomb structure. Now time was fixed by the monuments, rather than the monuments being built at the right time. Cultural change such as the emergence of an agricultural lifestyle allows for the development of new kinds of technology. This development then in turn creates new lifestyles for individuals and this in turn encourages further new forms of cultural change.

Thus the Neolithic people were able to differentiate time into what could be described as 'daily time', whereby they were living their ordinary lives, and 'cosmic time', which involved the monuments and a wider view of their world encompassing the past. This past involved myths which accounted for the formation of the Earth, but at the same time the people were aware of the future and knew that their megaliths would far outlive them.

It is possible therefore to distinguish three different but connected kinds of time from a human, lived perspective:

- The time of our lives, personal existence, bounded by life and death (living and dying);
- Public life, to do with marking important events or occasions and redolent with symbols and metaphors which must be resonant with daily lives if they are to have an effect;
- The enduring quality of long-term time, the concept of time that is embedded in tradition, ways of doing things and long-term continuities.

To keep time synchronous, because the 12 months as measured by the cycles of the Moon did not equal the one year determined by one cycle of the Sun through the sky, an extra month had to be inserted into calendars to equalize the lunar and solar cycles. This insertion marks the origin of humanity's material engagement with time as a shared cultural resource. As society became increasingly complex the marking of time evolved at different levels to facilitate this complexity in the form of periodic religious festivals for example, or on a smaller scale the regulation of day-to-day events such as the opening and closing of markets or the timing required for processes such as pottery firing and metal smelting. Thus time was being used to regulate daily life in these societies, but was also being used to develop a grander scale of time measurement to explain the possible origins of the universe or cosmos.

Rationalism and myth

The Greeks were among the earliest to engage rationally with time. As described earlier in this chapter, the observations of Herodotus and the statements of Heraclitus and the philosophical offerings from Aristotle suggested the idea of a changeless or cycling eternity or infinity.

Rationalism and myth each defined a response to time and the universe. The mythic response was to propose a universe born with time – that is the universe is formed and time begins with that formation. Before this creation there is no time, no duration and no ordered sequence of events. The story of the Creation in Genesis in the Old Testament falls into this category. In contrast the no-creation myths envisage a universe without a beginning. Some proposed a cosmos with an infinite past and an infinite future that would therefore need no beginning. Another possibility examined was a universe of eternal cycles where time is simply a loop played over and over again throughout eternity. The Hindu mythologies of the dance of Shiva fall into this category. Here endless cycles of creation and destruction are controlled by the choreography of the great god Shiva.

Creation myths imagine time as a straight line, while no-creation myths consider time either as a straight line extending infinitely in both directions or as a circle. We shall return to this dual behaviour of time in a later chapter.

As civilization developed and people began to live increasingly urban lives, the need for greater subdivision and recording of time became necessary and the first calendars were developed and the first attempts made to break the day down into smaller units. There is abundant archaeological evidence from cuneiform tablets to show that in the Mesopotamia region of the Middle East about 5 kya there were accurate calendars based on lunar cycles and the solar year to service the agricultural, economic and political needs of the population.

The tyranny of the clock

The subdivision of the day into regular smaller units began in the Middle Ages in the monasteries of Europe. Although the division of the day into 24 equal hours was a Babylonian invention and may be linked to the division of the Zodiac into 12 constellations, it was the desire for order in their daily worship that led the monasteries to develop the concept of regular subdivisions of the day and night. Prayer times were ordered by seven periods: matins, prime, terce, sext, none, vespers and compline. Matins began with sunrise, sext was approximately noon, none mid-afternoon, vespers was the end of the working day and evening prayers were set for compline. While the monks may have been motivated more by a desire for order in their daily prayers rather than a need for an accurate subdivision of time, there is no question that what began as an ordering of their prayer rituals became an ordering of daily life both inside and outside the monasteries.

By the end of the 14th century many cities in Europe were sporting a conspicuous mechanical clock with a prominent face showing the hour and bells sounding the time with loud peals. Thus was the day divided, never again to be undivided, as the increasing complexity of society required greater accuracy in the management of time. These initial clocks had only an hour hand as smaller divisions of time were not yet required, but the human experience of time had been entirely redesigned by the invention and spread of the clock. The advent of the minute hand in the late 17th century and the arrival of the Industrial Revolution (Fig. 2.11) in the late 18th century further compounded these changes.

The Industrial Revolution is the name given to the period from around 1760 to the middle of the 19th century when there was a transition to new manufacturing processes with a change to machine production, leading to major changes in society, including greater urbanization. This was in the context of an increasing percentage of the population leaving the land to work in a factory environment, governed by a rigid adherence to a timetable based not only on hours but on minutes. Until then most people were aware of the passage of time in only very general terms – seasonal changes, the phases of the Moon, the difference between night and day. All of this changed profoundly with the arrival of industrialization.

© PHB.cz(Richard Semik)/ Shutterstock

Figure 2.11 The Iron Bridge, Shropshire, symbol of the Industrial Revolution in England.

As with the revolution of the spread of agriculture during the Neolithic, here was an example of profound cultural and social change being driven by a change in the material culture of society, in this case the rise in urbanization inherent in the Industrial Revolution. In the interim period between the development of public clocks (and those belonging to factories) and the universal appearance of cheap and reliable clocks (alarm clocks in particular) an important role was played in industrial areas of Britain and Ireland by the 'knocker-upper'. This was a man or woman whose job it was to rap with a long pole on the bedroom windows of mine or mill workers to enable them to arrive at work to beat the strict timekeeping rules that were a feature of industrialized society by this time. Many mills and factories would lock their doors to late-coming workers who would thus lose their wages for that day, while repeated transgression would quickly result in dismissal. Here was a good example of the lives of working people being governed by calculation of time at the level of individual minutes while showing a discrepancy between the precision of the measurement of time available to the factory compared to that of its workers.

The rigorous timekeeping that typified the mills, mines and factories of the Industrial Revolution in Europe and North America was followed by the imposition of timetables on a wide range of activities – schools, commerce, government offices, hospitals – and an important driver of this process was public transport networks. In Britain stage coaches were operating to a published schedule in the late 18th century and by the 1840s railway timetables were common. In Britain the railway companies introduced a standard time for the country – railway time – and by 1880 the government had passed legislation standardizing time for the whole country and abolishing all local time zones which were based on solar time. For the first time in history the population of a country was required to live according to an artificially derived time frame. This situation is now taken to an extreme in China, which covers more than 50 degrees of longitude and has a single time zone for the whole country.

All of this manipulation of time was the precursor to the often frenetic life-style practised by many people in the developed world today, based on a 24-hour day with every hour subdivided into 60 minutes.

Our perception of time has changed since our Neolithic ancestors made the first tentative steps to measure and record time as part of their expanding culture. We now have the capability as never before to investigate time past and time in the future. Time as an enigma, as a straight line or a cycle, still exerts an enormous influence on how we live our lives and on how we view our past and contemplate our future.

We have a greater understanding of the origin of time and the origin of the universe than ever before, yet there is still a powerful lobby in the most advanced cultures in the developed world that believes the Earth is about 5,000 years old and was built over a period of 7 days. Discussions about the origin of the universe and what happened in the first micro-seconds after the 'Big Bang', and whether or not there even was such an event are unlikely to provide definitive answers for that part of the population outside those researching in the field of cosmology.

At the present state of knowledge, however, this book aims to investigate that segment of time that has passed since the origin of the Earth within the Solar System. To understand the history of the Earth has been the goal of generations of geologists over many hundreds of years, from even before the term geologist was invented.

So let us for the purposes of this study consider time as a succession of events, those events being placed in the order of their happening, with regard to their duration in time and affirming the continuance of Earth processes.

Chapter 3

The importance of understanding time

We see no vestige of a beginning, no prospect of an end…
James Hutton (1726–97), *Theory of the Earth*, 1788

The concept of deep time, the idea of geological time representing a past billions of years before our relatively recent human culture, was introduced in the previous chapter. There is a need to examine this fundamental principle of geology and show why it is vitally important for the future of the planet that it is better understood by society than has previously been the case. The gradual realization of the likely immensity of geological time from the 18th century onwards has proved to have major ramifications across a whole series of fields, not just science and religion but also in the realms of politics and economics. The impacts on science and religion are well documented and accounted for many of the most bitter controversies of the 18th and 19th centuries as geology developed as a separate science. In the 20th and 21st centuries it has become clear that the understanding of the environmental and economic challenges faced by society today cannot be properly understood without an appreciation of the concept of deep time. Problems such as the role of fossil fuels in society and their possible implication in climate change or the consequences of current rates of resource consumption by an increasing world population currently occupy world governments. I am convinced these dilemmas can only be dealt with satisfactorily if there is a greater appreciation within society of the timescale that governs the various cycles and processes that provide us with the resources that we are now so dependent on.

Despite being responsible for one of the great ideas that changed how we relate to our place in the world around us, geology as a profession does not appear to have done enough to ensure that society in general appreciates the extent of geological time. There cannot be many people who believe today that the Earth is flat, or that the Sun orbits the Earth, but there are many who believe we live on a planet that is less than 10 kyr old. As late as 1845 the English poet John William Burgon was describing the ancient Jordanian city of Petra as 'a rose-red city half as old as time'. This is a more accurate description of the red sandstone out of which these magnificent buildings are carved, than of the city's antiquity. The Nabataeans occupied this site not more than 3 kya, long after the pyramids in Egypt were built.

The Scientific Revolution that began in the Middle Ages had greater impact in its spatial rather than its temporal aspects. This is coupled with an inability among the public to identify the order of major events in Earth history and a tendency to conflate events which were in fact widely separated in time. A good example of this is the widely perceived co-existence of dinosaurs and humans. Surveys carried out among college students in the USA (Libarkin *et al.*, 2007) show that while many can place in order events such as the origin of life, the extinction of the dinosaurs and the rise of *Homo sapiens*, there is a poor sense of the actual amount of time existing between these events.

A major impediment to the greater appreciation of deep time in geology is widespread religious teaching that is resistant to the idea of an Earth of great antiquity. A recent poll in the USA (Gallup, 2010) found that 40% of the population believe that humans were created in their present form some 10 kya. To put this period into the context of geological history, our own species, *Homo sapiens*, is believed to have developed nearly 200 kya. Around 10 kya the major ice sheets were melting and retreating from the northern hemisphere at the end of the last Ice Age, leading to the development and spread of agriculture.

It appears that the extent of geological time poses a fundamental problem to many people who have difficulty coming to terms with the apparently insignificant period that we have been on the planet. Humans not only have been present for a very short period but also appear to play a minor role in the wider picture of the universe. In addition, what appears to be a lack of purpose in the occurrence of natural events throughout geological time is philosophically unsettling for many people. This is in contrast to, for example, the Christian doctrine which provides a meaning and rationale for life. This perceived lack of meaning in the modern geological narrative perhaps leads to reluctance among these groups to engage with the whole concept of deep time.

It is worrying that there appears to be such widespread indifference in society in general to the significance of geological time. The implications of deep time are too important to be left in the hands of the scientific community and such insights would better inform people when making the sorts of decisions that will have to be made if society is to remain viable in the 21st century and beyond. One way to awaken a greater appreciation of time is to place it within a wider societal context, by persuading people that a better understanding of deep time will, for example, enhance their appreciation of the surrounding landscape. In addition an appreciation of geological time would help in the understanding of the basis of sustainable development using renewable energy rather than non-renewable fossil fuels. In this context the significance of long-term climate changes becomes more than an abstract concept argued over by politicians and scientists.

So, why should a non-geologist take an interest in what is clearly a difficult concept to grasp? What possible relevance is there for society as a whole in something that appears to be closely associated with men digging up fossil dinosaur bones in the deserts of Utah or China? In the view of many members of the public, geologists are considered to be a mildly eccentric group of specialists in an obscure subject. They are wheeled out periodically to explain an exceptional earthquake (nowadays often linked to a potential tsunami), a volcanic eruption or the latest oil or gas find in a hitherto pristine part of the world. Society generally fails to realize the extent to which civilization, however it is defined, depends on the knowledge and skill of the geological community to go on finding the materials required for the level of infrastructure and domestic technology regarded as essential in our everyday lives. These materials include not just exotic commodities like gold and diamonds but also more mundane materials such as iron and copper and the ingredients for the vast amounts of concrete poured every day all over the world. It is not an exaggeration to state that without rocks we would never have developed socially, culturally or industrially and the old geological adage – if it cannot be grown it has to be mined or quarried – is as relevant as ever.

The Bingham Canyon copper mine, located near Salt Lake City, Utah, USA, is one of the largest open-pit mines in the world, measuring over 4 km wide and 1,200 m deep (Fig. 3.1). The mine exploits what is known as a porphyry copper deposit where a crystal-rich magma has moved upwards through the crust. Hot fluids circulated through the magma and surrounding rocks and deposited copper-bearing minerals to form the deposit. About half a million tonnes of rock are removed for processing each day. The time taken to take out the copper, probably around 100 years, is a small fraction of the time taken to form the deposit in the first place. Porphyry copper deposits are part of a crustal cycle that operates over millions of years. This means that resources such as these are considered 'non-renewable'. Copper deposits such as Bingham Canyon are feeding the ever-increasing demand for copper in the world as electrification and the demand for domestic appliances rises. The potential economic and environmental impacts of resource exploitation on this sort of scale can be better assessed in the light of an understanding of the geological processes and timescale associated with their origin.

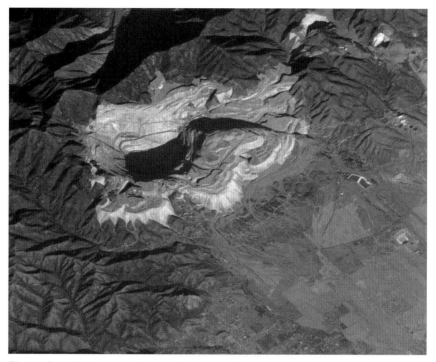

Figure 3.1 The Bingham Canyon mine, viewed from space. NASA image.

In fact, the concept of deep time is one of the most culturally relevant ideas in the history of human development. If the varied temporal spans are examined – for example the period since the end of the last Ice Age, the length of time required to build and erode a mountain range, the time span for the opening and closing of an ocean, or the rate at which species evolve – they offer perspectives from economic, political and cultural viewpoints.

Economic implications of geological time

There are important economic implications of geological time, particularly in the field of resource supplies. Society today relies on a wide range of natural materials that result from processes operating over very long periods and often under physical and chemical conditions that are difficult if not impossible to reproduce. The production of fossil fuels such as coal, oil and gas from decaying organic matter takes place over many millions of years and because of this the rate at which we are using these materials is not sustainable. The formation of important metal deposits such as copper or zinc is associated with processes deep in the crust which also take place over hundreds of millions of years. In contrast a resource such as topsoil can be formed in a time frame of the order of hundreds or thousands of years rather than millions, but against that the

Figure 3.2 An abandoned farm in South Dakota during the Dust Bowl conditions of the 1930s. USDA image.

rate of soil erosion under certain climatic and farming conditions can be much quicker. The Dust Bowl was a period of severe dust storms affecting the agriculture of the US and Canadian prairies, particularly the states of Oklahoma and South Dakota, in the 1930s. In the drive to increase the area under cultivation, deep ploughing had displaced the native deep-rooted grasses leading to conditions that allowed the soil to dry out and be blown away by the prevailing winds (Fig. 3.2).

In many areas affected by the Dust Bowl conditions there has still not been a restoration of soil conditions, even after a period of nearly 80 years. The Oklahoma Dust Bowl is a 20th-century example of a failure to recognize the complexities of the grassland ecology of the Great Plains, and a further failure to appreciate the amount of time that would be required to regenerate the soil layers which had blown away.

The water cycle (Fig. 3.3), whereby water evaporates from the oceans, is carried by clouds to the land where it falls as rain and returns via surface water and ground water to the oceans to be recycled, operates on a timescale of a few hundreds or thousands of years, depending on the physical scale of the cycle.

Water is stored underground in aquifers, permeable bodies of rock, often on a vast scale. In many parts of the world these aquifers are an important

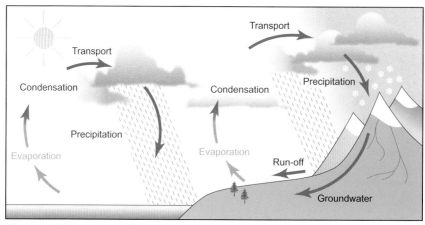

Figure 3.3 The water cycle.

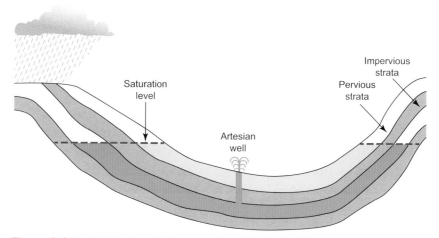

Figure 3.4 Aquifer showing artesian properties, after Andrew Dunn.

and sometimes the only source of water. Figure 3.4 shows a simple aquifer of layers of pervious (permeable) rock, sandwiched between layers of impervious (impermeable) rock. Water can pass through permeable rocks but not through impermeable rocks. The water enters the system from the edges and travels underground through the permeable layers, filling the aquifer to the saturation level. If the aquifer is tapped near its centre the head of water may exert sufficient pressure to force the water to the surface without the requirement of pumping and these conditions are described as artesian. The distance across such an artesian basin may be many hundreds of kilometres and in many parts of the world, the Sahara Desert for example, artesian basins are a vital source of water. The length of time water spends in the groundwater portion of the water cycle is called the residence time and may be as little as days or 10,000 years or more.

The residence time can be calculated by a variety of methods, including the measurement of the rate of flow of water through the aquifer and the use of tracers to follow flow paths and estimate storage capacities.

It is essential to recognize, however, that if consumption of water from any aquifer, whatever the scale, is greater than the rate of renewal, then the amount of water held in the aquifer will steadily decline. The Ogallala Aquifer is part of the High Plains Aquifer System in the mid-western states of the USA and is one of the biggest in the world occupying an area of around 40,000 square kilometres (km^2). It is a major source of irrigation water in those states such as Nebraska, Kansas and Texas that it underlies. Monitoring of the capacity of the aquifer shows that water is being depleted by extraction rates that exceed the recharge rates in a number of areas where the amount of water stored is declining. The rate of depletion is accelerating and clearly this is a cause for great concern given the significance of the aquifer to the overall water supply of the USA. Although the water cycle is relatively short compared with some other geological cycles, sustainable management of any aquifer requires an understanding of the timescale of water infiltration and replenishment rates in the system. World demand for water will steadily increase as living standards gradually improve and the need for careful management of water as a resource will become even more important than the management of oil and gas resources. We can live without using petroleum products, and we did so until relatively recently in our history, but we cannot survive without water.

Political implications of geological time

By political implications of geological time I mean the potential relationship between deep time and political decision-making. Probably many politicians and other decision-makers would be puzzled at being asked about such an influence, as would many of the people who voted for them, but once again this is a reflection of the reluctance of the general public to embrace the concept of deep time.

Two themes are dominant in the future planning by most governments today, certainly those of the developed world. These are:

- sustainable use of resources including energy supplies and strategic minerals;
- the existence or otherwise of human-derived climate change and its implications.

A feeling for the timescales involved in the formation of the so-called non-renewable resources such as oil, gas and metal ores is essential for decision-making involving the use and conservation of these products which are vital to the world we now find ourselves in. Similarly decisions on climate change can only be sensibly made with an understanding of what has happened to the Earth's climate, not only millions or billions of years ago, but also in the recent past, over the last 100,000 years and especially over the last 10,000 years as the ice melted and retreated and the Earth entered its current interglacial period. Society has developed an increasing awareness of the roles and responsibilities of science and the subsequent need for scientists of all disciplines to inform the public of the consequences of their research. There is a requirement for geoscientists to inform decisions by looking at the geological past and by extrapolation to look into the geological future.

Cultural implications of geological time: the landscape as a cultural item

The idea of 'landscape tourism' is a relatively recent phenomenon, but does mark an awareness change in the view people have of their environment. Many coastal communities have realized that having people visit to watch whales in their natural habitat is vastly more profitable than killing the whales for food. In the same way certain regions are becoming more aware of the attractiveness of their landscape features to tourists. The idea of visiting an area such as the Grand Canyon solely to look at a geological phenomenon may appear to be an aesthetic choice, typical perhaps of societies developed enough to have sufficient leisure time and disposable income for self-indulgence. It should be remembered, however, that our earliest ancestors, across all continents, were capable of recognizing features of their landscapes which were special. These special areas were often important culturally and were then physically enhanced in various ways, for example the building of megalithic monuments on prominent sites across western Europe in the Neolithic.

John McPhee, who we met in the previous chapters, examined the concept of deep time on a journey across the basin and range country between Utah and California in the USA. Travelling with a geologist companion, McPhee passed through vanished landscapes represented by the ancient rocks seen in the present-day surface and often examined by the pair in road cuttings. His geologist companion regarded these as windows looking into the past. Using a different analogy, a road-cutting to a geologist is like a stethoscope to a physician – a way of gaining information about the interior of a body. In the case of the geologist the body in question is the Earth. To read the story of the landscape in this way, as described earlier when examining the hidden landscapes of north-west Scotland, is to appreciate our surroundings at a higher level of understanding than if we just drive through it and wonder at the spectacular scenery.

The consumer society

A more mundane aspect of the cultural context of deep time concerns the rise of what is commonly known as the 'consumer society'. This is a focus by the developed countries on the acquisition of consumer goods as epitomized by motor vehicles, electrical appliances and complex communication networks. All of these require large amounts of resources in the form of metals such as steel, copper and aluminium as well as huge quantities of lower value materials such as crushed rock aggregates, sands and gravels. All of these projects consume energy, much of which is some form or other of fossil fuel – coal, oil or gas. An appreciation of the timescales required to produce these resources quickly leads to the conclusion that these resources are non-renewable within the lifetime of human endeavour. Countries such as China and India a few decades ago were routinely described as 'underdeveloped'. They are now fast becoming leading world economies and between them account for around 37% of the world's population. If their populations aspire to the same living standards as the USA or the European Union, and the current global economy certainly encourages this, then the consumer society as we have known it is finished. The Earth simply does not have the reserves of petroleum products or ores of strategic elements to service such an expansion. A major cultural change in what society views as important will be required.

What all of this means is that while geoscientists of all types may see the need for the rest of society to have a greater understanding and appreciation of the concept of deep time, they have to realize that they have yet to convince society of this. This failure to engage has resulted in a lack of knowledge of and apathy to deep time among the population as a whole. More effort has to be applied to using thinking about time past as a guide to thinking about time future if we are to have a sustainable future on this planet.

We need now to look at those pioneers of chronology whose vision, imagination, observational skills and laboratory expertise allowed at least a partial unravelling of the story of the passage of geological time on Earth.

Chapter 4

The early chronologers

The Bible shows the way to heaven, not the way the heavens go…
Attributed to Galileo Galilei (1564–1642)

Understanding geological time is not an easy task. While the concept of an ancient, perhaps infinite, Earth had been considered by early Greek and Roman thinkers, the philosophical conflict between the supporters of deep time and those who believe in a young Earth was established at an early stage and continues to this day. The 'discovery' of the 'New World' by Christopher Columbus in 1492 was one of a number of happenings around the end of the 15th century and the beginning of the 16th century that began the challenges to many of the orthodoxies of science that had emanated from the classical world. These challenges can be regarded as a revolution in scientific thought, and the willingness by a number of courageous visionaries to question previously held views led to a prolonged period of progress in various fields of science. Included among these visionaries were scientists and philosophers who were interested in the study of time. They can be thought of as the earliest chronologers – from the Greek *chronos* meaning time.

The Scientific Revolution

By the Middle Ages a variety of models had been put forward to explain the origin and age of the Earth within the diverse cultures and societies across the world. Many of these stories were of great antiquity, often showing overlap with each other in themes and details and evoking a higher authority of some kind, from the monotheistic god of the Judeo-Christian and Muslim traditions, to the polytheism invoked by Hindus. The problem faced by these early societies was to explain the inexplicable. We are fortunate that there have been major scientific advances in the last 500 years which allow us greater insight into the processes that brought about the origin of the Earth. However, when it comes to explaining the birth of the universe and what preceded it and what will succeed it, many people would feel we are not really much wiser than our ancestors.

In Britain and the rest of Europe the influence of the Bible, principally the Old Testament, was dominant in philosophical thought and anyone contemplating the origin of the world was constrained by the account given in Genesis. This said the world was made in 6 days and at a subsequent stage there was a worldwide flood involving Noah and his family. The Gilgamesh flood myth of the ancient Babylonians is another example of many cultures across the world where there is a folk-memory of a deluge as part of their creation story and the Gilgamesh epic displays many features in common with the later Genesis version. However, by the 16th century in Europe, developments in science – the onset of the so-called 'Scientific Revolution' – began to challenge the religious orthodoxy of the times, leading to the overthrow of many of the ideas held as sacrosanct by the Christian churches. This was a religion versus science debate that in some senses is still taking place. The dichotomy in thought and philosophy that had been established in Classical Greek times between pragmatists like Herodotus who recognized the likely origin of the Nile Delta as formed by successive layers of sediment deposited over tens of thousands of years, and those like Heraclitus and Aristotle who thought that time was changeless and endless (see Chapter 2), was mirrored in the Middle Ages. Scientists like Copernicus and Galileo were seeking to change the Earth-centred view of the universe by putting the Sun at the centre of the Solar System, something the Church considered as heresy.

Following the advances made in mathematics, astronomy and the understanding of the cosmos by the early Greek philosophers, the culmination had been the geocentric scheme of celestial motion of Ptolemy (90–168 CE) whereby the planets moved around the Earth in circular orbits.

Ptolemy's model (see Fig. 4.1), published as *Syntaxus Mathematica* in 140 CE, was the standard interpretation for the next 1,000 years, carried on through the Dark Ages in Europe by Arabic astronomers. By the middle of the 16th century, however, things began to change. In 1543, Nicolaus Copernicus, just before he died, published his book *On the Revolutions of the Celestial Spheres*, considered a major event in the history of science as it proposed a model for the Solar System that placed the Sun at the centre. This meant the Earth was in orbit around the Sun and not the other way around. Following on from this pioneering heliocentric model in the mid-16th century, by the early 17th century the revolution in scientific thought was fully under way. Long-established ideas were challenged by progressives of the scientific world such as Galileo Galilei, René Descartes and Johannes Kepler.

Galileo Galilei (Fig. 4.2) was an Italian scientist, born in 1564 near Pisa, the son of Vincenzo Galilei who was a lutenist, composer and musicologist. Vincenzo was a major figure in the musical life at the beginning of the Baroque

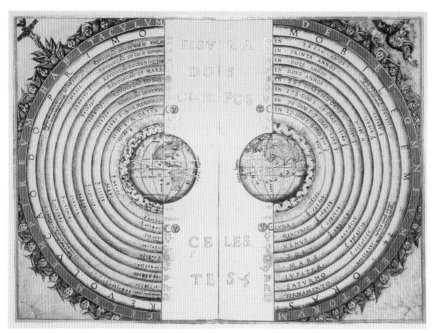

Figure 4.1 Ptolemy's model of the universe, this version is by the Portuguese cosmographer Bartolomeu Velho, 1568.

Figure 4.2 Portrait of Galileo (1564–1642) by Justus Sustermans.

era. Later in his life he studied the mathematics of pitch and string tension and made some important discoveries in the field of acoustics. It is likely the career path of his son was influenced by his father's activities in mathematics and physics.

Galileo Galilei began academic life as a medical student at the University of Pisa but changed to mathematics and philosophy, eventually becoming professor of mathematics at Pisa and then Padua. He is probably best known to the public for his association with the leaning tower of the town of his birth. Here he investigated gravity by dropping objects of different weights from the upper stories and measuring their rate of fall, formulating the basic law of falling bodies. He also constructed a telescope with which he studied lunar craters and sunspots and discovered four moons revolving around Jupiter.

However, he was a supporter of the idea of Copernicus that the Earth was in orbit around the Sun and this was dangerous intellectual territory for anyone during these times. Since the Church regarded the holders of such beliefs as dissenters, Galileo found himself under investigation by the Roman Inquisition in 1615 – 'under vehement suspicion of heresy'. This was the first of a series of confrontations with the Church authorities, leading initially to a ban on teaching or advocating these theories. Eventually in 1633 he was sentenced to life imprisonment, reduced to permanent house arrest. He was also forced to publically recant his belief in the Copernican theory. Galileo spent the remainder of his life confined to his villa south of Florence and died in January 1642.

It was finally accepted he may have been right when in 1992 Pope John Paul II acknowledged that the Catholic Church had erred in condemning Galileo for asserting that the Earth revolves around the Sun. The 17th-century Church won a victory which can only be described as pyrrhic – one where victory is as costly as defeat. The Church won the battle but lost the war. In his years of house arrest Galileo's continued experimental investigations established scientific practices that would pave the way for future generations of physicists and other scientists. No less a scientist than Isaac Newton, born the year Galileo died, would come to use his findings when making his own contribution to the Scientific Revolution.

Overlapping with Galileo in the first half of the 17th century was the French philosopher and scientist René Descartes (Fig. 4.3), famous now for his observation – *cogito ergo sum* – I think therefore I am.

Descartes lived mostly in the Netherlands and was an important supporter of the heliocentric view of the Solar System espoused by Copernicus in the preceding century. Descartes took an anthropocentric view of life; he believed that human reason was autonomous and he rejected any appeals to divine ends to explain natural phenomena, preferring instead to account for physical phenomena by way of mechanical explanations. This was a further step in the separation of science from the churches and was a plank in the development of scientific thought in Europe. His book *Traité du monde et de la lumière* (*The World* or *Treatise on the Light*) was ready for publication in 1633 and in it he had adopted the same heliocentric model of the Solar System advocated by Copernicus and

Figure 4.3 Portrait of René Descartes (1596–1630) by Frans Hals.

later by Galileo. This, however, was the year of Galileo's final denunciation by the Inquisition and Descartes, fearing he would suffer the same fate as Galileo, decided against publication. *The World* consisted of several related works covering physics, mechanics, animals and man and was eventually published posthumously in 1664. His later book *Meditations*, published in 1647, parted company with the 'old' science based on the principles of Aristotle. In doing so he is viewed as establishing the ground for the changes in the fundamental understanding of science that began in Europe towards the end of the Renaissance era and continued through the late 18th century.

Johannes Kepler (Fig. 4.4) was another of those scientists who had been profoundly influenced by the work of Copernicus, and in 1596 he wrote an outspoken defence of the Copernican System.

A German by birth Kepler moved to Prague to work with the Danish astronomer Tycho Brahe. In 1609 he published *Astronomia nova* delineating his first two laws of planetary motion. In the book he says: 'I demonstrate by means of philosophy that the Earth is round, and is inhabited on all sides; that it is insignificantly small, and is borne through the stars'. For the times he lived in this was a bold statement indeed. Concentrating on the orbit of Mars, Kepler overthrew the dominance of the circle as the basis for explanations of planetary motion since the work of Aristotle and Ptolemy. The superb data gathered by his Danish mentor Tycho Brahe allowed him to realize that the orbit of Mars could in fact be explained by an ellipse and from this he implied that all planets move in elliptical orbits in a Sun-centred Solar System.

Astronomia nova was followed in 1621 by his *Epitome Astronomiae*, which was his most influential work and discussed all of heliocentric astronomy in a systematic way. In it he proposed his third law of planetary motion (the law of harmonies).

His first law (see Fig. 4.5) establishes that the orbit of a planet is an ellipse around the Sun. According to the second law (again see Fig. 4.5), a line from a planet to the Sun sweeps out equal areas (A) in equal times (t). Earth is therefore faster at the position perihelion, when it is closer to the Sun, and slower at aphelion, when it is farther from the Sun.

Figure 4.4 Portrait of Johannes Kepler (1571-1630). Artist unknown.

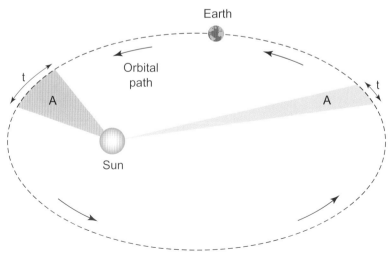

Figure 4.5 Kepler's first law of orbital ellipse and second law of areas

However, while Kepler can be considered as the first to correctly explain planetary motion and is therefore a key figure in the Scientific Revolution, he stuck to biblical orthodoxy when considering time and infinity. He believed the cosmic clock had started with the Creation and calculated the date 3992 BCE for the creation of the world – an age of less than 6,000 years. In this respect Kepler was, like so many of his contemporary scientists, prepared to challenge the orthodox views on the spatial structure and dimension of the universe, but unwilling to rock the Church's boat when it came to the origins of time and the cosmos. Given the fate of Galileo and other 'heretics' during this period this is perhaps an understandable reaction.

James Ussher and the year of the Creation

The response to this questioning of orthodoxy by those such as Kepler and Galileo was the publication of a large number of calculations of the age of the Earth based on biblical texts. Among these probably the most famous was one by James Ussher (Fig. 4.6), Archbishop of Armagh in Ireland.

He famously calculated the date of the Creation as the night preceding the 23rd day of October, 4004 BCE. Ussher was a considerable scholar and his biblical chronology, *Annales Veteris Testamanti* (Annals of the Old Testament, see Fig. 4.7), published in 1650, became the definitive biblical chronology of the world, and one that still has adherents today.

Figure 4.6 James Ussher, Archbishop of Armagh (1580-1656) after Sir Peter Lely.
© National Portrait Gallery, London

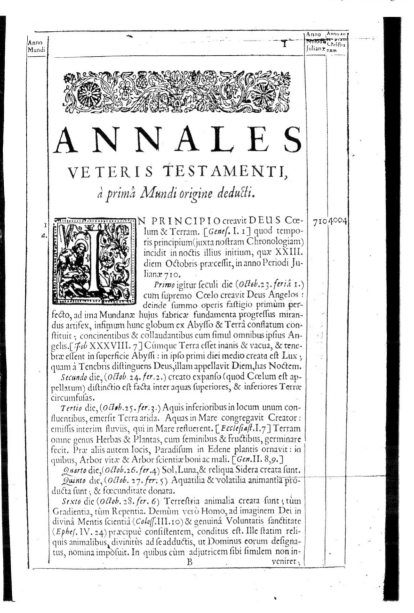

Figure 4.7 The first page of James Ussher's Annales Veteris Testamenti (1650). The date 4004 BC is included in the margin.

The first paragraph in Latin can be translated as: 'In the beginning God created Heaven and Earth, Which beginning of time, according to our chronologie, fell upon the entrance of the night preceding, the twenty third day of Octob[er], in the year of the Julian Calendar, 710'. The equivalent of 710 in the Julian Calendar is 4004 BCE.

41

In 1654 he added Part 2, which took his account of history through Rome's destruction of the Temple in Jerusalem in 70 CE. This comprehensive history of the world drew on Persian, Greek and Roman sources to create a history of the ancient world that is in remarkable agreement with modern versions.

The plethora of so-called sacred theories of the Earth that emerged in the 17th century were not exclusively concerned with the age of the Earth, but instead were attempts at explaining the nature of the Earth in the context of biblical events such as the Creation and the Flood.

Earlier in the century René Descartes, who had done much to initiate the separation of science from the influence of the Church, published a naturalistic model for the formation of the Earth and planets, *Principia philosophiae* (1644). He believed they were formed of material derived from an extinguished star, settling out in layers as it cooled. This can be considered a precursor of the nebular hypothesis for the formation of the Solar System, where a primitive solar nebula (a cloud of dust and gas) contained the building blocks of the Solar System as it exists today. Descartes attempted to explain the features of the Earth in terms of the physical and chemical processes he saw operating around him.

Thomas Burnet and his Sacred Theory of the Earth

Descartes' approach was in contrast to one of the best known models for the origin of the Earth, the Sacred Theory of the Earth (*Telluris Theoria Sacra*) published by Thomas Burnet, an Anglican minister, in 1681. While Burnet was undoubtedly influenced by the earlier ideas of Descartes, his model was one of a number which attempted to reconcile the Earth's origin with the events recorded in biblical texts. The major innovation of Burnet's work was an attempt to compute the volume of water present during Noah's Flood and to explain where all the water had come from to cover the Earth to a depth of 15 cubits (approximately 7 m). He suggested using normal sounding techniques for this, but seriously underestimated both the areal extent of the oceans and their average depth. As part of his overall model for the origin of the Earth he reasoned that the Earth had been smooth before the Flood and therefore less water would be required for such a coverage, and before the deluge the Earth had split open and the Flood was supplied from the Earth's internal water, this coming from a worldwide layer of water which was under the crust and concentric with it.

The seven globes shown on the title page of Burnet's First Edition (Fig. 4.8) each represent a stage in the evolution of the Earth. Stages 1 and 2 represent the early chaotic form of the Earth, out of which was created the paradise of Eden when the surface of the Earth was smooth and featureless. Stage 3 is the Flood, the aftermath of which produces the current surface of

Figure 4.8 Thomas Burnet — the title page of Telluris Theoria Sacra

the Earth. Stages 5, 6 and 7 represent the future, with 5 being the conflagration predicted in the Book of Revelation. After this is the Restoration in the Millennium when the Earth is restored to perfection, saints and sinners are separated and the Earth then becomes a star.

We noted in Chapter 2 that creation myths imagine time as a straight line in contrast to no-creation myths which consider time either as a straight line extending infinitely in both directions or as a circle. In many ways Burnet's model can be regarded as time as a straight line. Each stage of the process from creation to Earth becoming a star is a separate, non-repeatable event. Yet close examination of the model shows that the globes are arranged in a circle and the figure representing Jesus at the top straddles the beginning and the end and the translation of the Greek inscription around his head is 'I am the alpha and the omega'. Time is again seen as capable of dual behaviour.

Into the abyss of geological time

In the late 16th and early 17th centuries the changed perceptions of time and space, proposed by Kepler and Descartes among others, led to a backlash from the Church authorities and sanctions such as the house arrest imposed on Galileo. It is probably the case that by the late 17th and early 18th centuries an age of the Earth of around 6,000 years as proposed by Archbishop Ussher and a Bible-based creation story along the lines of Burnet's Sacred Theory could be considered as the conventional view. However, changes were under way and by the middle years of the 18th century a number of personalities were again pre-pared to challenge the churches and sought to separate science from religion in a new drive for rationalism.

One such naturalist/philosopher was Georges-Louis Leclerc, Compte de Buffon (1707–88). Convinced that the Earth had to be older than envisaged or allowed by the Church, in 1749 he published the first of more than 40 volumes of his book *Histoire Naturelle*. In this he suggested the Earth had formed as a result of a collision between the Sun and a passing comet, causing material to be ejected. He also saw the Earth's history as cyclical with no beginning or end, an idea put forward by James Hutton later in the century. The members of the Faculty of Theology at the University of the Sorbonne were less than impressed by these opinions and demanded a retraction on the grounds that the ideas were contrary to the creed of the Church. Buffon did so but with little sincerity and never stopped publishing the offending volumes.

In the 1760s Buffon began a series of experiments involving the heating of iron balls to near melting point and then measuring the time it took them to cool to a temperature where they would not burn the fingers. By extrapolation he then estimated how long the Earth would take to cool to the same temperature. He arrived at a figure of almost 50 kyr to cool to the point where fingers were safe and almost 97 kyr to fall to the present temperature of the Earth. Modifying his experiment by mixing metallic and non-metallic substances to more closely resemble the Earth's composition he modified his age estimate to 75 kyr and published it in 1778 in his book *Des Époques de la Nature*, probably the most famous of his scientific essays. An age such as he was proposing was of course more than could be countenanced by the Church authorities and Buffon's work had little impact at the time. However, his ideas on calculating the age of the

Earth by measuring its rate of cooling were to be pursued enthusiastically by William Thomson, the Lord Kelvin, nearly a hundred years later. In the meantime, at the time of Buffon's death in 1788, a young Earth based on the biblical account including a universal flood was still the orthodoxy.

A champion of this orthodoxy was Abraham Gottlob Werner (1749–1817), considered by many as the most influential geologist of the 18th century. Werner was a German geologist and mineralogist who spent most of his professional life in the mining industry in Saxony, employed as Inspector and Teacher of Mining at the Freiberg Mining Academy. He seems to have had an extraordinary talent for teaching. Students were attracted from all over Europe, and then spread his theories on the origin of rocks throughout their own countries. Werner's view of Earth history postulated a universal ocean which, having covered the entire globe, had gradually receded to its present level, precipitating all the rocks and minerals of the crust while doing so. He taught that all rocks, except lava flows clearly associated with active volcanic vents, were produced by precipitation or deposition from this all-covering ocean. Active volcanoes were considered to be the result of the combustion of underground coal beds which melted the overlying rocks to produce molten lavas. Differing rock types resulted from variations in the depth and other conditions in the universal ocean. Due to the association with the universal ocean adherents to this view were known as Neptunists, after the Roman god of the sea.

In the second half of the 18th century this view of the origin of the Earth and the formation of its rocks began to be challenged, notably by James Hutton (1726–97). Having visited the Scottish Highlands in Chapter 1 when we first considered the immensity of geological time, we return to Scotland to consider the significant contribution made by Hutton, a native of the city of Edinburgh, to our overall understanding of the concept of deep time.

The Scottish Enlightenment and James Hutton

The Scottish Enlightenment was the period in the 18th century when Scotland was characterized by a wide range of scientific and intellectual accomplishments. By the 1750s the Scottish population was among the most literate in Europe with a culture of debate and argument within its ancient universities and learned societies. A feature of many of the innovative thinkers of the period was a humanist and rationalist outlook on life and recognition of the fundamental importance of human reason. These were attitudes shared with the European Enlightenment occurring at the same time on the continent. Important figures in this progressive environment were the philosopher David Hume, Adam Smith who was the author of *Wealth of Nations*, considered by many as the first modern work of economics, and Robert Burns, the poet and lyricist. The contempt for the established authorities displayed by Burns in some of his works mirrors the changes in attitude beginning to be seen among those scientists and philosophers who were increasingly at odds with the Church doctrines of the time. It was against this intense intellectual background that James Hutton (Fig. 4.9) and his collaborator and biographer John Playfair conducted experiments and made field observations that were to have a profound effect on our subsequent appreciation of the concept of deep time and the processes whereby the Earth renewed itself.

Hutton was born in 1726, the son of a merchant who had served as Edinburgh City Treasurer, and began his university education in Edinburgh at age 14. Having moved from the study of the classics to medicine, he eventually completed a Doctor of Medicine degree at the University of Leyden in Holland in 1749. He became relatively wealthy after developing a successful business manufacturing the chemical sal ammoniac (ammonium chloride), used for dyeing and in a variety of other manufacturing processes. In the 1750s Hutton was engaged in agriculture on farmland in Berwickshire in Scotland, which he inherited from his father. During this time he was developing ideas in the fields of meteorology, mineralogy and geology as well as exploring agricultural practices from around Britain. It was in this period he began to recognize the importance of the processes of weathering and erosion and their relationship

Photograph by J Lyle.

Figure 4.9 The James Hutton Memorial Garden in Edinburgh on the site of his former home. Visible to the left of the plaque is a sample from Glen Tilt showing intrusive granite veins and a conglomerate boulder which has been carried by ice and water to signify cyclicity.

to the formation of new land areas. Returning to live in Edinburgh in 1767 he was soon an active participant in the intellectual life of the city and continued his geological studies. It was in this stimulating environment that he began to build his challenge to the Neptunist model for the origin of rocks and the age of the Earth.

By 1785 Hutton had written a paper, entitled 'Theory of the Earth: or an investigation of the laws observable in the composition, dissolution, and restoration of land upon the globe', read at the Royal Society of Edinburgh. In it he put forward his ideas on a perpetually renewing Earth that could be explained by understanding processes operating today. He was also beginning to recognize how vast geological time had to be to accommodate these processes. Hutton made a number of field observations to gather evidence in support of his theories. He did not believe that most rocks and minerals were precipitated from a universal ocean, after all they were not soluble in even very hot water. He surmised they could only become liquid if melted at very high temperatures. In 1785 he travelled to Glen Tilt in the Cairngorm Mountains in the Scottish Highlands to examine the relationship between granite and adjacent metamorphic rocks. Here he found veins of granite penetrating metamorphic rocks indicating that the granite had been molten at the time of coming into contact with the metamorphic rock, and the granite must also therefore be younger than the metamorphic rock.

In the years following Hutton's visit to Glen Tilt he received important backing for his views on the origin of igneous rocks from Sir James Hall (1761–1832) who was the first scientist to develop experimental research in geology. Hall was convinced that heat and pressure played an important role in the formation of igneous rocks and carried out a series of experiments to melt various rocks and minerals and cool them at a controlled rate. This helped confirm that the veins of granite observed by Hutton at Glen Tilt had been injected in a molten state into the overlying rocks. In addition Hall's experiments established the formation and composition of basalt lava and showed that if limestone was strongly heated under pressure it was changed to marble. The insistence that igneous rocks had a molten origin led to Hutton, Hall and their supporters being referred to either as Plutonists or Vulcanists, the implication being the rocks originated at depth in the Earth (Pluto was the Roman god of the underworld), or that they were formed by volcanic action.

Following his 1785 visit to Glen Tilt, in 1787 Hutton described what became known as Hutton's Unconformity at Jedburgh in south-east Scotland (Fig. 4.10). The exposure in the bank of the River Jed shows tilted sedimentary rocks (grey sandstones) overlain by horizontal sedimentary rocks (red sandstones). The junction between the two types of rock represents a period

of time when there was no deposition. Such a boundary is known as an unconformity and since there is such a marked difference in the angles made by the
beds (near-vertical below and near-horizontal above) the term angular unconformity can be applied. The unconformity shows that deposition is not continuous and can be interrupted by periods of erosion that follow tilting, folding
and uplift before sedimentation is renewed.

Figure 4.10 Hutton's Unconformity at Jedburgh. Photograph by Dave Souza.

An angular unconformity is a surface of erosion that separates two sets of beds where the orientations of the rocks are at different angles. The section shown above in Figure 4.10 can be interpreted in four steps as follows:

- A sequence of sediments is laid down horizontally.
- The sediments are buried and folded during a mountain-building event or orogeny.
- The mountains are uplifted and eroded producing a level surface or peneplain.
- After subsidence the second set of sediments is deposited horizontally, thus producing the angular unconformity.

This is also the situation we examined in north-west Scotland in Chapter 1, where the Torridonian sandstones are above the Lewisian gneiss peneplain.

Hutton continued to search for field evidence which would support his theories of a cycle of deposition, uplift and erosion over very long periods of time. In 1788 Hutton, along with his friends John Playfair and James Hall, found a sequence of sediments at Siccar Point on the Berwickshire coast east of Edinburgh (Fig. 4.11).

Here the greywacke beds (a form of sandstone) were aligned vertically. Above the greywacke layers were eroded fragments of greywacke which

Figure 4.11 The outcrop at Siccar Point, Berwickshire, showing sloping red sandstones overlying vertical grey sandstones

were in turn overlain by near-horizontal beds of reddish sandstone. Hutton correctly surmised the boundary between them, as at Jedburgh, represented a long time period between the first cycle of deposition followed by uplift and erosion before the next cycle of deposition. In fact we now know the greywacke rocks are about 425 myr old (Silurian) and the red sandstones are about 345 myr old (Devonian).

The interpretation was to make Hutton's name and move Playfair to exclaim famously (in a publication of 1805) 'The mind seemed to grow giddy by looking so far into the abyss of time' – surely one of the most evocative descriptions of the immensity of geological time ever expounded.

The word 'abyss' had been used in 1794 by Richard Kirwan in an attack on Hutton's ideas. Kirwan was a distinguished chemist and mineralogist and a leading Neptunist in Ireland and published a detailed counterblast to Hutton's ideas in the *Transactions of the Royal Irish Academy* in 1794. In the paper he said that not only was Hutton's theory contrary to the laws of Moses, but that it led 'to an abyss from which human reason recoils'. Kirwan's attack motivated Hutton, who had been suffering a period of ill health, to publish an expanded version of 'Theory of the Earth' in 1795 in which he reaffirmed his views and attempted to answer the criticism, both scientific and religious. To Kirwan he memorably retorted – 'The abyss from which the man of science should recoil is that of ignorance and error'.

One wonders whether Playfair's use of the phrase is with the benefit of hindsight some 11 years after Kirwan's comment and Hutton's response. In any event the minds of neither Hutton nor Playfair recoiled at the recognition of the immensity of geological time, and Hutton's place in the annals of the history of geological thought was assured.

Werner's legacy

Although the Neptunists eventually lost the argument to the Vulcanists, the contribution made by Werner to the development of geological science should not be underestimated. Among other things he recognized the importance of differentiating between sequences of rock units defined by their composition – lithostratigraphy – and those defined by their age – chronostratigraphy. In the latter case a particular rock type traced across the landscape may represent different times as conditions altered across a depositional basin. For example if the sea is gradually encroaching on a land area the beach sediment marking the sea boundary will become progressively younger as the sea covers more and more of the land surface. In the example shown below (Fig. 4.12) as the sea invades the land, a marine transgression, it deposits sands onshore, mudstones in deeper water and limestones in the deepest water offshore. The term onlap is used to

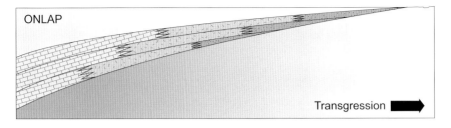

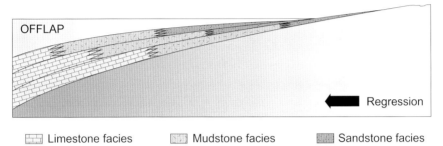

| ▭ Limestone facies | ▨ Mudstone facies | ▦ Sandstone facies |

Figure 4.12 Shifting sedimentary facies in the case of transgression (sea advancing) and regression (sea retreating). After Woudloper.

describe the process of successive beds extending on to older rocks as the transgression proceeds.

The term sedimentary facies is used to describe a distinctive rock type broadly corresponding to a certain environment or mode of origin. Any rock unit can pass laterally into a different facies and such a lateral change is termed a facies change, as illustrated above. The opposite case where sea level is falling is termed a marine regression. Here the term offlap is used to describe the process of progressively younger beds being deposited seawards.

This is an important concept and will be revisited when we deal with the correlation of rock units in later chapters, particularly the geology of the Grand Canyon.

Hutton's epilogue

Much has been said over the years about the poor writing style possessed by Hutton, and the dense prose that inhibited readers from engaging thoroughly with the ground-breaking ideas he was putting forward in his 'Theory of the Earth'. Further, his death in 1799 meant he was no longer able to promote his work. However, help was at hand. In John Playfair, his companion at Siccar Point and at other significant times in his career, he had someone who was prepared to play the role of James Boswell with Hutton as Dr Johnston, and to write a sympathetic biography. Playfair was an eminent scientist and mathematician in his own right, the Professor of Natural Philosophy at the University of

Edinburgh. In 1802 he published his now celebrated book entitled *Illustrations of the Huttonian Theory of the Earth*. This is widely recognized as extending the influence of Hutton's work posthumously. Hutton's insight had led him to believe that the Earth had a long history and that this history could be interpreted in terms of processes currently active on the Earth's surface. He proposed in effect that 'the future will resemble the past' and reasoned that there must have been numerous cycles of deposition and erosion enabling the Earth's surface to be renewed.

While Hutton, with the assistance of Playfair, undoubtedly made a major contribution to the development of geological science at the end of the 18th century, the real impact of his ideas came with their enthusiastic endorsement by probably the most eminent geologist in Britain in the first half of the next century, Sir Charles Lyell. Hutton's idea that processes currently operating on the Earth could explain Earth history became known as uniformitarianism and his phrase the future will resemble the past was changed to 'the present is the key to the past'. Lyell was now setting the scene for the next big controversy in the developing field of geology, uniformitarianism versus its antithesis catastrophism.

Chapter 5

Time's arrow and time's cycle

The present is the key to the past…
Charles Lyell (1797–1875), *Principles of Geology* 1830–33

In 1927 a British astronomer, Arthur Eddington, introduced the concept of the arrow of time or time's arrow, to explain the 'one-way direction' of time. About 60 years later, Stephen Jay Gould coupled time's arrow with time's cycle and published a book with the title incorporating the two terms. In viewing time as an arrow, each moment has its own distinct place in a sequence of events which are considered irreversible and unrepeatable.

Examples of important non-repeating events in Earth history include:

- change from an early oxygen-poor atmosphere to an oxygen-rich atmosphere;
- the apparent origin of life under conditions which no longer exist;
- evidence of larger and more frequent meteorite impacts in the past than today;
- the complete extinction of large groups of organisms throughout geological history – for example, the end-Permian extinction event which took place around 250 mya and wiped out more than 90% of species existing at the time.

According to time's arrow, history consists of a series of linked events moving in a particular direction. In contrast time can be viewed as a cycle. According to time's cycle, events have no meaning as distinct episodes and apparent motions are parts of repeating cycles and therefore time has no direction. History according to the Bible is primarily time's arrow – the Earth was created once, the commandments were delivered to Moses at a particular time and place, Noah's Flood was a unique event. There is, however, an overlay of time's cycle on this narrative, for example when cyclic events such as the solar cycle are invoked as metaphor – 'there is no new thing under the sun' (Ecclesiastes).

The terms uniformitarianism and catastrophism as used in geology owe their origin to a highly influential figure in 19th-century Britain, William Whewell (1794–1866). Whewell can truly be described as a polymath. He wrote extensively on many subjects, including mineralogy, geology, astronomy and the history and philosophy of science.

Many of the major scientists of the time, including Charles Darwin and Michael Faraday, turned to him for scientific advice, but his most lasting influence probably lies with his aptitude for terminology. The invention of the word 'scientist' is attributed to Whewell, as are the terms we are dealing with in this chapter, uniformitarianism and catastrophism.

Uniformitarianism is the assumption that the same natural laws and processes that operate in the universe now have always operated in the past. Catastrophism is the theory that the Earth has been affected during its history by sudden, short-lived events, often violent and with worldwide implications. Today there is a consensus that the rates of geological processes or their precise nature have not necessarily remained exactly the same throughout geological time; for example, volcanism may have been more frequent in the past. While today most geologists would accept a role for both catastrophist and uniformitarianist viewpoints, in the early 19th century philosophical standpoints on this debate were much more strongly polarized. The idea that Earth history consists of slow gradual change punctuated by occasional catastrophic events has taken until the latter half of the 20th century to be accepted as a conventional standpoint. The controversy at the start of the 19th century, as we shall see, had long-lasting implications for the development of geological thought in the 20th century.

There were two principal antagonists in this controversy. In Britain Sir Charles Lyell was the champion of the legacy of James Hutton and author of the highly influential textbook *Principles of Geology*. In France, meanwhile, Georges Cuvier recognized extinction episodes in the fossil record and explained these as the result of recurring natural events such as great floods and mountain-building episodes that punctuated long periods of stability during Earth history.

To understand the relationship between such contrasting philosophies in geology as uniformitarianism and catastrophism we must examine processes and events in Earth history which can be used to illustrate the two perspectives. Scientific advances in geology in the 20th century have enormously increased our understanding of the cycles and events which have shaped our planet. One area of particular progress has been in the understanding of the interrelatedness that binds many of those geological factors recognized by the early pioneers in the subject. No longer do those who study volcanoes work in isolation from those whose interest is in fossils or earthquakes or the processes of sedimentation in the oceans. Among geologists there has been a realization that each of those fields, and all the other specialist fields that have arisen in the last 50 years, can inform each other. The one thing they have in common is geological time.

One of the singular most important advantages scientists working today in fields such as geology, astronomy and cosmology have compared to their predecessors in the 19th and early 20th centuries is the ability to view the Earth and the surrounding universe from space, from outside the Earth's atmosphere. The pioneering photographs taken in the 1950s of the dark side of the Moon were the first items of data received on Earth from space. Nowadays there is a stream of such information, including data from routine monitoring of the Earth for a wide range of reasons, and we receive detailed information from our neighbours in the Solar System and beyond. Recently astonishing images of Mars have been beamed back from a robot traversing its surface and the Rosetta probe has just landed on the surface of a comet millions of kilometres from Earth in deep space. Aerial photography in geology has enabled detailed interpretations of many of the Earth's major features that would not have been possible otherwise. Let us now consider that part of the Earth which attains the greatest height above sea level – the mountains of the Himalayas and the adjacent Tibetan Plateau.

Figure 5.1 shows northern India and the snow-capped Himalayan mountain chain with the high Tibetan Plateau beyond the mountains to the north. The Himalayan Mountains are the eastern end of an almost continuous chain of mountains that extends from the European Alps in the west through the Caucasus Mountains in southern Russia and the Zagros Mountains in Iran.

The extent of the mountain range is marked by a zone of earthquakes trending roughly east–west from the western Mediterranean to the western borders of China, including the seismically very active Tibetan Plateau. This plateau is the largest and highest mountain plateau area on Earth, covering 2.5 million km^2 with an average elevation of more than 5 km and is estimated to be still rising at the rate of about 5 mm per year. The reason the Himalayas were formed and why the Tibetan Plateau is still rising is due to the mobile nature of the Earth's crust. Large segments of the crust are moving towards or away from each other and so the face of the Earth is slowly but constantly changing.

Plate tectonics

By the 1960s the science of geology had undergone a revolution. A radical idea, continental drift, whereby the continents moved or 'drifted' across the surface of the Earth, had been proposed in the early part of the century by

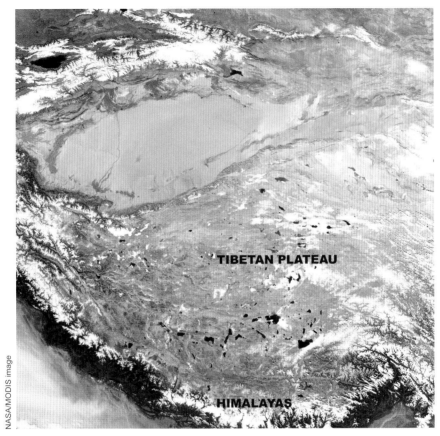

Figure 5.1 Northern India, the Himalayas and the Tibetan Plateau viewed from space.

Alfred Wegener, a German meteorologist. This idea, however, was considered to be heresy by the scientific establishment and did not find widespread acceptance. In the late 1950s and early 1960s new data, particularly from the ocean basins, led to a revival of the idea and its modification and expansion as the theory of plate tectonics.

The theory states that the surface of the Earth is not fixed but consists of about a dozen polygonal pieces called plates, which are made up of the rigid upper part of the mantle and the overlying continental or oceanic crust. Together the crust and upper mantle layers are referred to as lithosphere. The distribution of plates on the surface of the Earth (Fig. 5.2) resemble the fragments of cracked eggshell on a hard-boiled egg. These plates move in a variety of ways, producing different types of boundaries (Fig. 5.3):

- away from each other, as in oceanic spreading ridges such as the mid-Atlantic Ridge with the Americas plates moving away from the African and Eurasian plates (a divergent or constructive plate boundary);

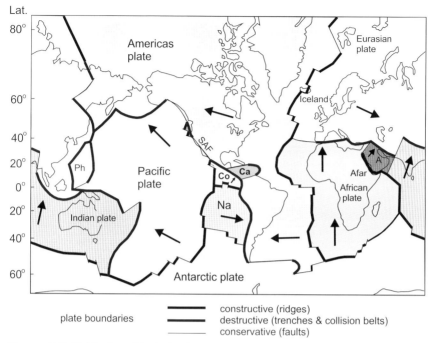

Figure 5.2 Distribution of tectonic plates on the Earth's surface and their boundaries. The plates are separated by three types of boundary: constructive – ocean ridges; destructive – ocean trenches; and conservative – faults. Small plates: Na, Nazca; Co, Cocos; Ca, Caribbean; Ph, Philippine; A, Arabian. SAF, San Andreas fault. The arrows give the direction of motion of each plate relative to the Antarctic plate. After Park.

- past each other, as on the west coast of North America (a conservative plate boundary marked by a transform or strike-slip fault);
- towards each other, as on the western coast of South America (a convergent or destructive plate boundary).

The rate of movement is around 5–10 centimetres (cm) per year and results from upwelling convection currents in the underlying mantle zone of the Earth's interior (Fig. 5.4). Convection currents form in liquids when heated. The hotter material becomes lighter and rises and on cooling becomes heavier and sinks. This sets up a circulation in the liquid. Although the mantle is solid it is sufficiently plastic to support convection cells. The rising convection currents at the mid-ocean ridges push the plates apart, while at the trench the plate is forced back into the mantle by the force of gravity pulling the plate downwards, exerting what is known as slab pull. This means that through time the distribution of continents and oceans on the surface of the Earth is subject to slow change.

Where an oceanic and a continental plate are moving towards each other oceanic crust is destroyed as it is forced under the continental crust. These zones are associated with major earthquakes and explosive volcanoes such as

Figure 5.3 Types of plate boundaries. After USGS.

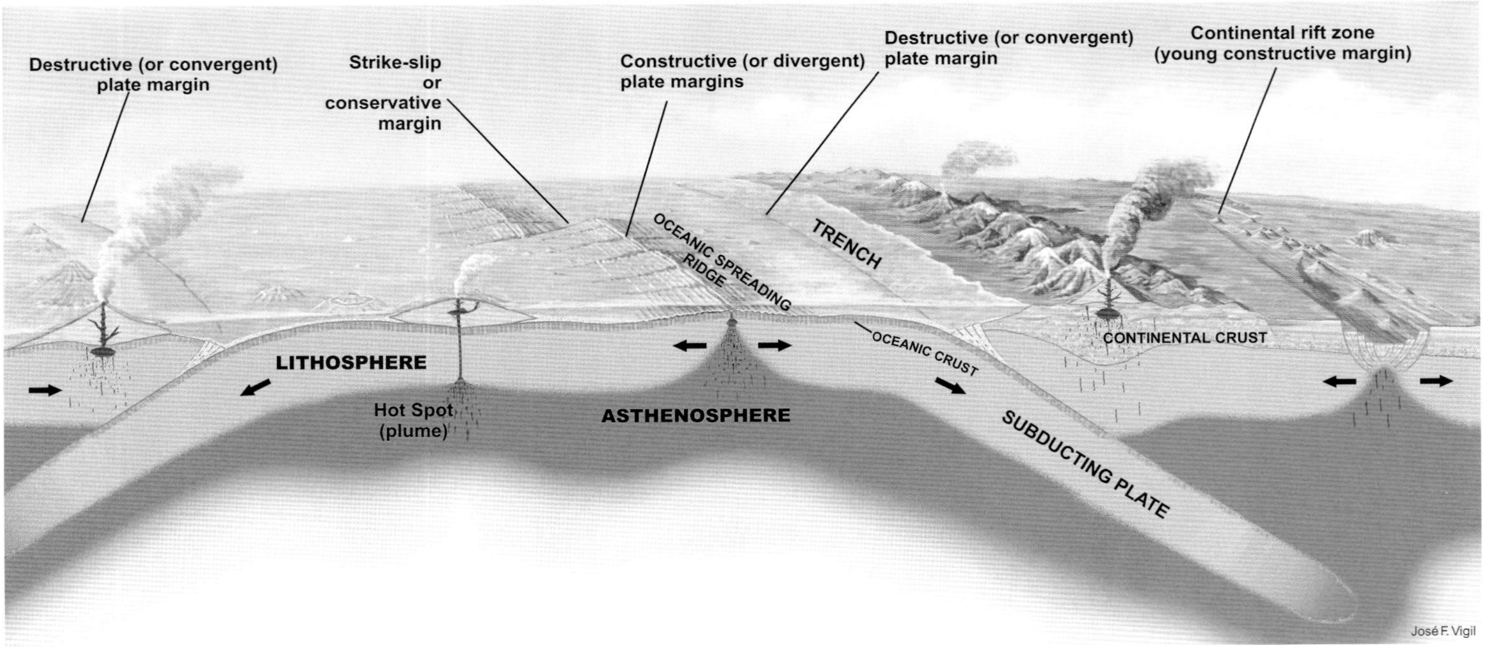

Continental rift zone
(young constructive margin)

Destructive (or convergent)
plate margin

Constructive (or divergent)
plate margins

Strike-slip
or
conservative
margin

Destructive (or convergent)
plate margin

TRENCH

OCEANIC SPREADING
RIDGE

CONTINENTAL CRUST

OCEANIC CRUST

SUBDUCTING PLATE

ASTHENOSPHERE

LITHOSPHERE

Hot Spot
(plume)

José F. Vigil

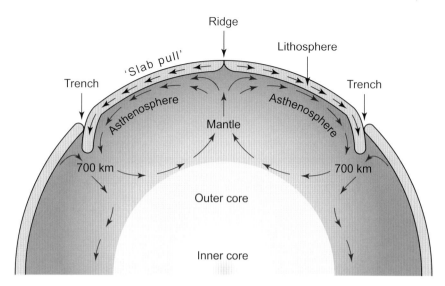

Figure 5.4 Convection cells in the Earth's mantle. After USGS.

the Andes chain in South America. Where two continental plates are forced together the crust is uplifted to form major mountain ranges such as the Himalayas. They formed when India moved northwards at about 15 cm per year and collided with Asia about 60 mya (Fig. 5.5). This movement is still continuing with the result that the Tibetan Plateau to the north of India is still rising at the rate of a few millimetres per year.

The Wilson Cycle

The Tibetan Plateau is a major feature of the Earth's crust and much of our understanding of its origin comes from a remarkable Canadian geologist, John Tuzo Wilson, whose contribution to geology in the latter half of the 20th century was immense. In the early days of the development of the plate tectonic theory there was a concentration on processes operating in the oceans, partly because much of the new data collected was the result of research into ocean floor characteristics. Tuzo Wilson's genius was to bring the continents more into consideration by recognizing the significance of the process of continental collision. He realized that if ocean basins are formed by the rifting apart of continents, then elsewhere on the Earth's surface other oceans must be closing, with continents colliding to form mountain ranges.

While recognizing that the Atlantic is steadily widening at the moment, Wilson asked, in 1966, whether the Atlantic had closed at some earlier stage and then reopened. In answer to his own question he proposed a pre-Atlantic Ocean called Iapetus which opened about 600 mya and closed about 420 mya.

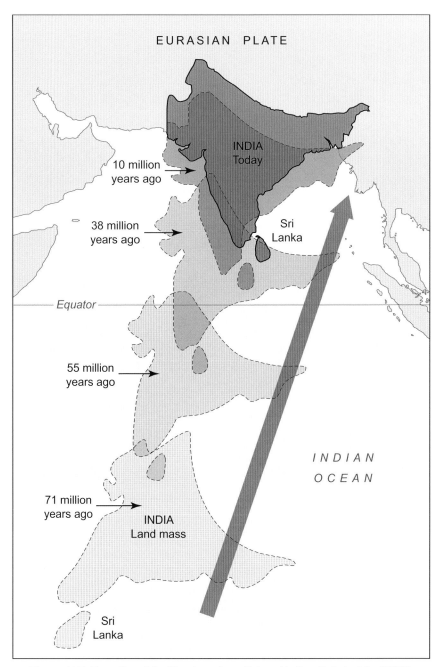

Figure 5.5 Formation of the Himalayas by collision of India with Asia. After USGS.

The birth and death of the Iapetus Ocean

Around 1,000 mya the continents on the Earth's surface had amalgamated to form a supercontinent called Rodinia. Over a period from around 750 mya to 550 mya this supercontinent broke up to form a new ocean with the newly formed continents Laurentia and Gondwana drifting apart as the ocean widened. Since Laurentia contains much of modern-day North America and on the other side of the rift were fragments of continental crust recognizable as modern-day northern Europe, Baltica and Avalonia, the ocean was a forerunner of the Atlantic. Accordingly it was named the Iapetus Ocean, after Iapetus, the father of Atlas in Greek mythology and after whom the Atlantic was named.

After about 100 myr, around 480 mya, the ocean had reached its maximum width and began to contract. The continental masses began a process of collision with each other and as the ocean floor shrank it was compressed as if in a vast vice. Laurentia, Baltica and Avalonia drifted closer to each other over a period of about 60 myr so that by about 420 mya the Iapetus Ocean was completely closed (Fig. 5.6).

The main consequence of the closure of Iapetus was the mountain-building event, the Caledonian orogeny, which formed the great Caledonian mountain range (the Caledonides) which stretched from Norway across Scotland and Ireland and in an unbroken chain into Newfoundland, eastern Canada and the Appalachians of eastern USA.

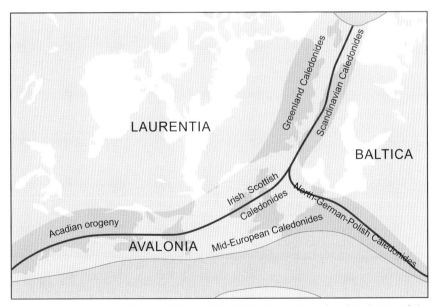

Figure 5.6 Reconstruction of America, Greenland and Europe after the closure of the Iapetus Ocean and the formation of the Caledonian mountain chain around 420 mya. After Woudloper.

One other consequence of this continental collision was to lay the geological foundations of Britain and Ireland as we know it today. The closure brought about the juxtaposition of the northern parts of Britain and Ireland which were part of Laurentia and the southern parts which were part of Avalonia.

This explanation of the processes involved in the opening and closing of Iapetus allowed the recognition of one of the largest scale cycles operating on Earth, the Wilson Cycle. The Wilson Cycle (Fig. 5.7) consists of a number of clearly defined phases:

1. **The embryonic phase**: A previously stable area of continental crust, a craton, is split as rising convection currents cause large volumes of melted mantle known as magma to rise and heat the crust, thereby causing it to stretch and rift (e.g. the East African Rift).
2. **The juvenile phase:** Sea-floor spreading starts and a juvenile ocean opens (e.g. the Red Sea–Gulf of Aden area).
3. **The mature phase:** A mature ocean basin develops (e.g. the Atlantic Ocean).
4. **The declining phase:** Progressive closure of ocean basins occurs by subduction of ocean lithosphere after the ocean basin reaches its maximum width (e.g. the Pacific Ocean).
5. **The terminal phase:** Shrinking of the ocean basin occurs and continental collision begins (e.g. the Mediterranean area).
6. **The suturing phase:** With the complete closure of the ocean basin the two continental margins of the ocean are joined or sutured together. They continue to collide and form an uplifted mountain range (e.g. the Himalayas and the Tibetan Plateau). This mountain range is subject to erosion, leading to a flattened or peneplained surface.

The operation of the Wilson Cycle takes hundreds of millions of years for the full range of events to be realized. This means it is the most imposing of all the variations of time's cycle examined so far. It represents the most transformational cycle of all the recurring events evident on Earth.

The supercontinent cycle

The Wilson Cycle operates on a timescale of a few hundred million years. Iapetus began to open perhaps 600 mya and had closed by 420 mya, around 200 myr for the formation and closure of an ocean basin. The present South Atlantic began to open about 150 mya. By around 60 mya the North Atlantic was forming and this ocean is still expanding.

However, the Atlantic is approaching maturity and relatively soon in geological terms it could develop a subduction zone on either its western or eastern

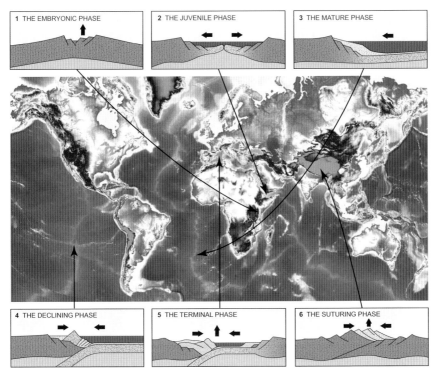

Figure 5.7 Stages of the Wilson Cycle showing the formation, development and closure of an ocean basin. After Hannes Grobe.

margin (the declining phase in the cycle) and the ocean will begin to close. When the Iapetus Ocean closed around 420 mya it heralded the eventual amalgamation of all the continental areas on the Earth's surface into a single supercontinent called Pangaea (meaning all land). Pangaea is one of a number of such continental agglomerations that can be recognized as having occurred throughout geological time. The break-up of the supercontinent Rodinia has already been cited as the precursor to the opening of Iapetus. It seems that periods when continents amalgamate are followed by periods of continental dispersal, at intervals of a few hundred million years.

Pangaea formed around 270 mya when it occupied nearly one-third of the Earth's surface. The distribution of continents today is the result of the break-up of Pangaea (Fig. 5.8), first into the two major land masses Laurasia and Gondwana, and then further fragmentation to give the pattern of the modern world we are familiar with. This periodic coming together and dispersal of the Earth's continents can be described as the supercontinent cycle and while there are varying opinions as to whether the amount of continental crust is increasing, decreasing or staying about the same, it is agreed that the Earth>s crust is constantly being reconfigured.

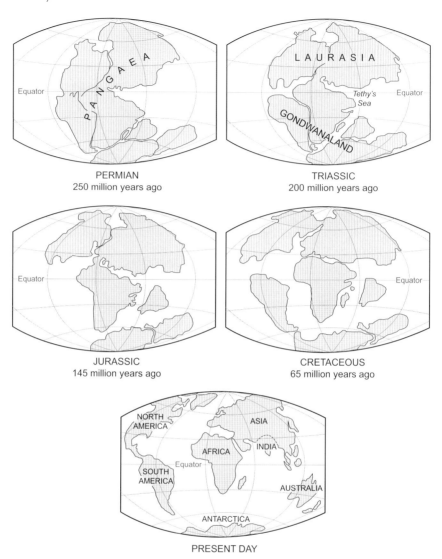

PERMIAN
250 million years ago

TRIASSIC
200 million years ago

JURASSIC
145 million years ago

CRETACEOUS
65 million years ago

PRESENT DAY

Figure 5.8 The break-up of Pangaea. After USGS.

The supercontinent cycle is the extension of the Wilson Cycle and complements it. Since the oldest sea floor is around 200 myr old, and the most ancient bit of continental crust goes back to 4 byr or more, it makes sense to emphasize the much longer record of cyclic events that is recorded in the continents. We shall return to the supercontinent cycle later in the book (Chapter 9) when dealing with future time.

However, in our continuing examination of time's cycle we should now look at one of the most iconic geological features on the Earth's surface – The Grand Canyon.

The Grand Canyon of the Colorado River: riding time's cycle

The Grand Canyon serves as an illustration of Hutton's views on uniformitarianism. He recognized the significance of unconformities such as that at Siccar Point, which represented crustal uplift followed by erosion and deposition of sediments. This was, he thought, a renewal of the Earth by the same processes that had always operated throughout geological time. This principle of renewal can clearly be demonstrated in the walls of the Grand Canyon. Starting at the base of the succession and working upwards, the frequent episodes of uplift, erosion and sedimentation with advances and retreats of the seas are laid out as in the pages of a book, though in this case the book has to be read vertically, going from bottom to top and from oldest to youngest.

It is difficult to imagine the impact created by the view (Fig. 5.9) when first seen by Europeans (probably in the 16th century), and even more so to imagine how the first geologists reacted when they came upon it in the mid-19th century. Those of us fortunate enough to have visited this most remarkable geological phenomenon have almost certainly been prepared for the scale of the canyon by some form of photographic image prior to visiting it for the first time. However, even with this foreknowledge there can be few vistas in the natural world that can compare with the Grand Canyon of the Colorado River. The scale of the width and depth of the canyon (29 km and 1,800 m respectively) can barely be appreciated from this image taken from the South Rim, and to recognize the full length of the chasm (446 km) requires a view from space (Fig. 5.10).

Figure 5.9 Grand Canyon of the Colorado River. © Gert Hochmuth/Shutterstock

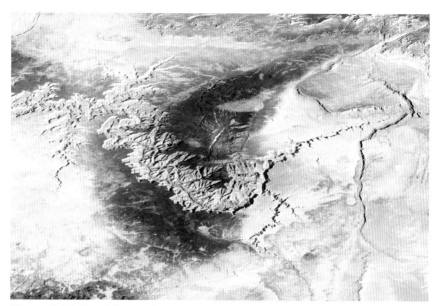

Figure 5.10 Grand Canyon from space. The view is looking from the South Rim to the North Rim, centre field. NASA image.

For the 17 myr since the Colorado River established its course it has incised downwards, cutting across the Colorado Plateau as the plateau was simultaneously uplifted, so enhancing the erosional effect of the river and producing a landscape image clearly visible from space. One of the intriguing features of the Colorado Plateau is its stability. Like the Tibetan Plateau discussed earlier it is the product of plate tectonic processes over a long time span but it shows little by way of rock deformation (folding and faulting) within the last 600 myr or so. This means the layers of sedimentary rock are close to their original horizontal orientation. In contrast, regions around the Colorado Plateau such as the Rocky Mountains to the north and east were thrust up as part of a mountain-building event such as formed the Himalayas. The Laramide Orogeny started around 75 mya, caused by subduction processes off the western coast of North America, and about 18 mya the Basin and Range Province to the west and south of the Colorado Plateau formed as the Earth's crust was subjected to stretching and cracking. Throughout these events the Colorado Plateau was raised over 3 km but still remained more or less horizontal.

While the three dimensions of the Grand Canyon (length, breadth and depth) are impressive, it is the fourth dimension that is truly awe-inspiring. That fourth dimension of course is time. From the bottom of the canyon, where the river is visible, to the canyon rim it is estimated that there is a difference in age of 1,250 myr. That is 1,250 myr out of the 4,500 myr since the Earth formed in the early days of the Solar System, an appreciable segment of geological time. Such an opportunity to examine the progress of time's cycle over nearly one-third of Earth's history does not present itself anywhere else on the surface of the planet, so in that sense the Grand Canyon is unique.

The sequence of rocks in the Grand Canyon

In Chapter 1 the principle of 'deep time' was discussed with the idea that a series of rock layers laid down in sequence could be perceived as a 'timeline' and read as a sequence of geological events. The rocks of the Colorado Plateau have many stories to tell: of the advance and retreat of seas and oceans over the area; of the arrival and departure of a variety of organisms as life evolved in increasing complexity; and of the range of marine and terrestrial environments in which these sediments were deposited and in which these organisms lived and died.

A pioneer in the exploration of the Grand Canyon was John Wesley Powell, a one-armed American Civil War veteran. In 1869 he led an expedition down the Colorado River in what was then completely uncharted territory (Fig. 5.11). After an epic journey of 99 days, which not all members of the expedition survived, Powell emerged to become one of America's pioneering heroes. In the years following his trail-blazing exploration of the Canyon he led several geological and ethnological expeditions to Arizona and Utah, becoming Director of the US Geological Survey in 1881. Although the 1869 expedition created a sensation with the public, it produced little of scientific value. The later, government-sponsored, trip in 1871–72 produced photographs and detailed maps.

Photograph by E. O. Beaman [Public domain], via Wikimedia Commons

Figure 5.11 First camp of the second John Wesley Powell expedition, Green River, Wyoming Territory.

Even a casual glance at the walls of the Grand Canyon will show differences in the various layers making up the sequence, or succession. Parts of the walls are steep-sided cliffs, while other parts are less steep and appear to consist of loose material or rock debris in the form of screes which hide the bare rock underneath. This difference reflects the variation in hardness between the more resistant sandstones and limestones forming the cliffs or obvious steps in the canyon profile and the softer shales and mudstones which make up the succession between the more durable beds. The Bright Angel Shale near the bottom of the canyon is a good example of one of the more easily eroded sediments in the wall, while Kaibab Limestone forms the prominent layer at the rim of the canyon.

In a sequence of sedimentary rocks, those at the bottom of the sequence are older than those at the top. At the level of the river it is clear there are further differences in the appearance of the rocks. Instead of generally buff-coloured sandstones or pale limestones which show a clear near-horizontal layering, these basement rocks are dark in colour and appear to be steeply inclined.

This is the Vishnu Formation, consisting of metamorphic rocks called schists which have also been intruded by igneous rocks like granite. These are the oldest rocks found in the canyon and have been dated as being around 1,500 myr old and belong to the Precambrian. Immediately above the Vishnu is a group of sandstones and limestones, clearly younger than the Vishnu because they are above the schists, but tilted away from their original horizontal position (Fig. 5.12). These are the rocks of the Grand Canyon Supergroup and are around 600 myr old and are therefore also part of the Precambrian. The

Figure 5.12 Dark coloured Vishnu schists at the base of the sequence in the Grand Canyon with the paler coloured Grand Canyon beds above them, forming an angular unconformity. Photograph courtesy of R. Smylie

boundary between these two groups of rocks is marked by a great difference in their ages (1,500 and 600 myr), but also by a change in the orientation of the beds, the younger rocks are at a different angle. This is an angular unconformity, precisely the same relationship recognized by James Hutton at Siccar Point and Jedburgh, and described in the previous chapter.

Above the Grand Canyon Supergroup rocks are the first beds of the largely horizontal sediments which make up the main part of the canyon walls. These are the Tapeat Sandstones and since they lie at a lower angle than the Grand Canyon Supergroup this boundary is also an unconformity, named the Great Unconformity by Powell in 1869.

Angular unconformities such as these not only record intervals of erosion but also provide a record of ancient earth movements. The beds below such unconformities were tilted, faulted, folded and uplifted before erosion produced the more or less even unconformable surfaces we see today. Unconformities represent renewal of the processes of erosion and deposition, while crustal uplift and mountain-building creates new exposures of rock which are then subjected to erosion leading to the deposition of fresh sediments, and so the cycle continues. Above the Great Unconformity it is possible to distinguish a further five major unconformities (Fig. 5.13). All of these represent a gap in time, but they also represent differences in the environment of deposition of the rocks, often associated with changes in sea level and therefore changes in the flora and fauna found in sediments.

Environments of deposition in the Grand Canyon

Of the rocks present in the Grand Canyon, we have so far looked mainly at the two oldest groups, the Vishnu basement rocks (metamorphic and igneous) and the Grand Canyon Supergroup composed of mainly sandstones and limestones. Above these rocks, across the Great Unconformity, are the layered Palaeozoic rocks. Palaeozoic means ancient life forms, and the term is applied to rocks which are older than about 250 myr but younger than about 540 myr. These rocks are mainly sandstones, limestones and shales and form some of the most conspicuous layers of the canyon. The oldest, the Tapeat Sandstones, are just over 500 myr old, while the youngest (the Kaibab Limestones) at the canyon rim are about 270 myr old. It is worth noting that the rocks in the deep parts of the Inner Canyon cover a range of around 1 byr up to the Great Unconformity, while the more spectacular visible part extends over a mere 230 myr.

The first of the Palaeozoic rocks, the Tapeat Sandstones, formed on an ancient sea shore, and they pass up into the Bright Angel Shale, a hardened marine mudstone containing invertebrate fossils known as trilobites, animals related to modern crayfish. Trilobites are useful fossils because they changed fairly rapidly

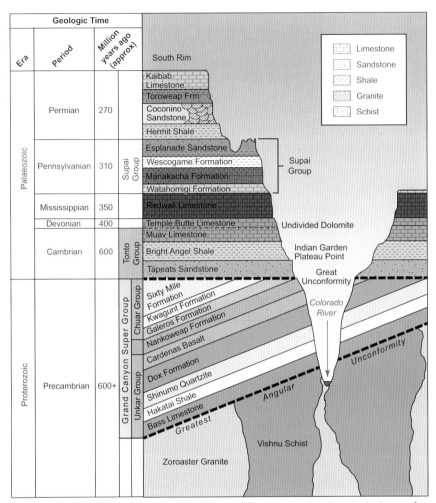

Figure 5.13 Geological succession at the Grand Canyon showing the principal unconformities. After USGS.

with time, so therefore changes in their shape and form can be used to work out the relative age of the rocks they are found in. (Relative age – whether the rocks are older or younger than other rocks – and absolute age – where the age of the rock is known in years – are discussed further in Chapters 6 and 7.) Examination of the trilobites found in the Bright Angel Shale showed that the rock in the west of the canyon is older than the same layer to the east. This means that the sea which deposited the Bright Angel Shale flooded the land surface to the east at a later stage. In other words as the sea was encroaching on the land surface from west to east in a marine transgression it continuously laid down sand along beaches and mud in deeper water. Thus sandstone will change laterally into mudstone. This is a facies change, a concept discussed earlier in Chapter 4.

Transgression and the opposite process, regression, imply vertical move-ments of sea level relative to the land surface and these movements can be due to local or regional folding and faulting, or they may reflect larger scale global changes in sea level. The expanding ocean which deposited the Bright Angel Shale continued its eastward progress and laid down the Muav Limestone above the shale as the water depth increased.

Above the Muav Limestone, and representing a break of about 65 myr, is the Temple Butte Limestone which was deposited on the eroded surface of the Muav. It is not known whether sediments were deposited during these 65 myr and were later removed by erosion or if they were never deposited in the first place. In either case the boundary between the two formations is another unconformity. The significance of the Temple Butte Limestone is that it con-tains vertebrate fossils, animals with backbones, unlike the trilobites in the older formations below. These animals are marine fish in the west of the canyon and freshwater fish in the east. The arrival of vertebrate marine fossils in the succes-sion is a significant milestone in the evolution of life on the planet and preceded the invasion of the land by amphibious life forms.

Unconformably above the Temple Butte Limestone with an age gap of around 20 myr is the prominent-cliff-forming Redwall Limestone. The environment in which the Redwall was laid down was a shallow tropical sea around 300 mya and the rock contains marine fossils such as corals and trilobites. This environment is in contrast to the overlying Supai Group, which consists of mud, sand and limestones deposited on a broad coastal plain similar to the Texas Gulf Coast of today. The eastern part yields fossil land plants and footprints of amphibians and reptiles showing that by this time life has evolved from fish through amphibians to reptiles and that both animal and plant life are now established on land.

An unconformity of about 5 myr marks the junction between the top of the Supai Group and the overlying Hermit Shale, although the Hermit Shale was laid down in a similar coastal plain environment and contains fossils of plants and ferns as well as vertebrate animal tracks. Unconformable above these beds is the Coconino Sandstone, formed about 275 mya as the region gradually became a desert. The Coconino beds are characterized by a particular form of layering called cross-bedding, which is characteristic of sand dunes (Fig. 5.14).

In the case of cross-bedding the inclined surfaces are not due to tilting or folding after deposition. Instead it is an indication of deposition by a flowing medium such as water or wind, with the deposition taking place on inclined sur-faces such as ripples and dunes. Desert sand grains have a very rounded appear-ance under the microscope (Fig. 5.15). Because of the constant attrition of being knocked against each other as they are blown about by the desert winds, the originally sharp-angled quartz grains are gradually worn into nearly perfect

Figure 5.14 Cross-bedded sandstone, Arizona.

1.0 mm

Figure 5.15 Desert sand grains showing a high degree of rounding due to wind abrasion.

spheres. The cross-bedding of the sandstone and the very rounded nature of the sand grains in the sediment confirms the desert environment of deposition of the Coconino beds.

Above yet another unconformity the Coconino Sandstone is overlain by the Toroweap Formation consisting of sandstones and interleaved limestones. These rocks were deposited in a sea that was advancing and retreating over the land and as the sea eventually deepened it formed the Kaibab Limestone about 270 mya. This is the uppermost and therefore youngest bed in the canyon walls and the one you are most likely to be standing on for your ceremonial photograph overlooking the depths of the canyon.

Most of the treatment of the geology of the Grand Canyon so far has related to time's cycle, with its emphasis on periodic renewal and uniformity of process. The fossil record of the canyon, however, simultaneously clearly shows time as an arrow. The evolution of life forms evidenced by the fossils in the canyon walls shows the development from invertebrates such as trilobites being the dominant form of life in the seas, to the arrival of marine vertebrates such as the earliest fishes. Following this were the amphibians and reptiles as animal life moved to the land. As this was happening plants were also beginning to establish themselves on land. Evolution is irreversible; time here is acting in a straight line – time's arrow.

According to Stephen Jay Gould, time's arrow consists of distinct and irreversible events, while time's cycle represents timeless order and law-like structure. This apparent dual attribute of time is one of the paradoxes that geologists have to wrestle with, but both characteristics are required to fully comprehend Earth history. As mentioned in the previous chapter, following on from the ideas of Hutton and uniformitarianism, the eminent Victorian geologist Sir Charles Lyell coined the phrase 'The present is the key to the past', meaning that past processes could only be explained by invoking processes which were the same as those observable today. Lyell's wholehearted embrace of uniformitarianism, coupled with a reaction in 19th-century Britain against the influence of the various churches on scientific thought, led to a fairly extreme interpretation of the idea, often to the detriment of objectivity. This was best illustrated by his clash with the French scientist Georges Cuvier, who believed that Earth history could be affected by short-term high-impact events – clearly a greater emphasis on time's arrow. Cuvier's ideas were totally rejected by Lyell and the geological establishment in Britain. By the middle of the 19th century therefore the scene was set for the next philosophical clash, time's cycle versus time's arrow, or Lyell versus Cuvier, a dispute that had far-reaching consequences into the middle of the 20th century.

The role of catastrophism in Earth history: time's arrow

Catastrophism is the theory that Earth history has been influenced in the past by sudden, short-lived, violent events which may have had worldwide implications. This view of course is in direct opposition to the idea of uniformitarianism, the dominant philosophy in geology for many years, which holds that slow incremental changes are responsible for Earth processes. The modern consensus on catastrophism is that catastrophic forces have indeed had a profound influence on episodes of Earth history, on both geological and biological processes. Examples of such events are collisions with large objects from space, asteroids, comets or meteorites for example. In 1994 there was indisputable evidence in the shape of Comet P/Shoemaker-Levy of the occurrence of such collisions within the Solar System.

From 16 July 1994 for around 6 days, pieces of Comet P/Shoemaker-Levy collided with the planet Jupiter, in the first collision of Solar System bodies ever to be observed. At least 21 fragments of the comet up to 2 km in diameter impacted on the planet, resulting in spectacular effects on Jupiter's atmosphere. This high-profile collision made the general public aware, if they had not been previously, that the Solar System, although seemingly vast, contained within it large bodies which could collide with Earth with devastating consequences for all life forms.

We need only look at the surface of our nearest neighbour in the Solar System, the Moon, to see evidence in the form of impact craters (Fig. 5.16) of the constant bombardment by meteorites and asteroids that the larger bodies in the Solar System have been subjected to.

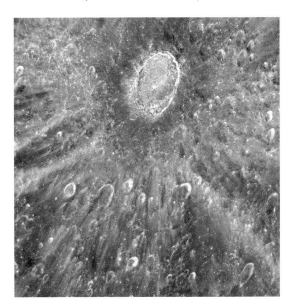

Figure 5.16 The Tycho impact crater on the moon.
NASA/ESA/ D.Ehrenreich

The early history of the Earth would have seen a much higher rate of collision as the young planet attracted Solar System debris by the force of gravity, growing steadily bigger as a result. While the likelihood of such a collision is much less now the probability still remains that the Earth's surface could be hit by an extraterrestrial body of sufficient size to cause catastrophic damage on a worldwide scale. Meteor Crater in Arizona (Fig. 5.17) resulted from the impact of a nickel-iron meteorite, probably about 25 m in diameter about 50,000 years ago. The resulting crater is about 1.2 km in diameter. It was the first such impact crater to be recognized on the surface of the Earth.

Since this recognition of the existence of impact craters on the surface of the Earth many more examples have been found, including probably the best known impact, the crater at Chicxulub on the Yucatan Peninsula in the Gulf of Mexico (Fig. 5.18). This impact is thought by many to be responsible for the demise of the dinosaurs. The crater is more than 180 km in diameter making it one of the largest impact structures on Earth.

The meteorite that caused the crater is estimated to have been at least 10 km in diameter. From the age of the surrounding rocks the impact occurred at the end of the Cretaceous Period, about 66 mya. Irrespective of whether or not the Chicxulub meteorite impact was directly responsible for the extinction of the dinosaurs, the worldwide effects of a collision on this scale would have been devastating. The colossal shock waves generated would have caused earthquakes and volcanic eruptions and these would have led to huge tsunamis

Figure 5.17 Meteor Crater, Arizona. © Vladislav Gajic/ Shutterstock.

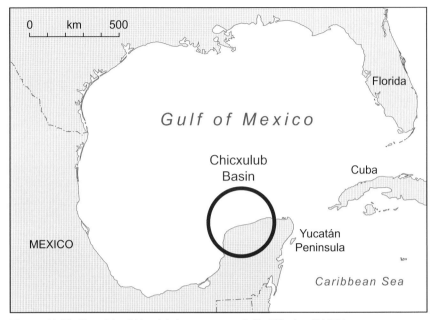

Figure 5.18 Site of the Chicxulub impact crater, Gulf of Mexico. NASA image.

across the world's oceans and vast clouds of super-heated dust and ash spreading from the impact site. In the longer term there would have been an increased greenhouse effect due to the release of vast quantities of carbon dioxide from the vaporization of limestone bedrock at the site of impact leading to greatly increased atmospheric temperatures. At the same time blockage of sunlight caused by dust and ash particles in the atmosphere would have disrupted plant photosynthesis which would in turn affect the whole food chain.

The Chicxulub impact event is a prime example of a seemingly random event affecting the Earth, something that is hardly predictable in anything but very general terms, but yet may have profound effects, on a worldwide scale. These would impact not only on physical aspects of the Earth but also on how species are developed or even whether or not they survive at all. While the reasons for the disappearance of the dinosaurs continues to grip the popular imagination, there are in fact many 'mass extinctions' discernible in the fossil record, of which this end-Cretaceous event is just one. Occurrences such as at Chicxulub, however, can be described as a catastrophe – defined as 'a violent convulsion of the globe, producing changes in the relative extent of land or water'. To underline the ongoing importance of catastrophic events in Earth's history, it is only necessary to remember two events in the last 100 years or so.

In 1908 near the remote Siberian locality of Tunguska an asteroid or comet crashed to Earth. Around 80 million trees were flattened (Fig. 5.19) over an area

Figure 5.19 Trees felled at Tunguska, 1908. NASA image.

of about 2,000 km² and the impact produced a shock estimated at 5.0 on the Richter Scale – the equivalent of a substantial earthquake. Amazingly there are no reports of serious injury or death associated with the event due mainly to the extreme remoteness of the area. In other circumstances an explosion of this magnitude would inflict major damage on a metropolitan area and its population.

More recently, in 2013, the Chelyabinsk Meteor exploded in the air over the Ural region of Russia. Estimated as 17–20 m in size, it weighed 10,000 tonnes and produced a bright fireball and small fragmentary meteorites. It generated a powerful shock wave which caused many injuries from flying glass and other debris. While both these events are small in comparison with Chicxulub, they both had the potential to cause major loss of life if they had struck in more densely populated areas of the globe.

Uniformitarianism versus catastrophism – Lyell against Cuvier

Following on from the challenges to the authority of the churches in the 18th century there had developed a widening gap between those scientists who believed in a rational explanation of Earth history and those who still adhered to a greater or less degree to the account of the origin of the Earth furnished by the Bible. Hutton had recognized the importance of cycles of erosion and deposition in the formation and development of the Earth. This advance, coupled with his realization of the immensity of geological time, meant that the beginning of the 19th century was in many ways the start of a new era in the study of the Earth. Hutton had shown that the history contained within successions of rocks could be unravelled and thus the observer could effectively travel back in time and view ancient, long-disappeared landscapes. For the first time it became possible to systematically understand the relationships between the various

Figure 5.20 Left; Sir Charles Lyell (1797–1832); Right: Georges Cuvier (1769–1832).

factors influencing the geology of a particular region. This involved the connection between climate and flora and fauna, the relationship between oceans and mountain belts and the role of volcanoes in Earth history.

In the early years of the 19th century Charles Lyell (Fig. 5.20) built on Hutton's ideas of cyclicity in Earth's processes. In doing so he gathered much evidence in support of the idea that the features of the Earth were the result of the same geological processes that could be observed currently, acting slowly over very long periods of time. His major work, *Principles of Geology*, was published in three volumes between 1830 and 1833 and was to be hugely influential in geology then and for many years after.

Lyell's view that uniformitarianism was the explanation of Earth processes was in direct contrast to the views of the French scientist Georges Cuvier. Cuvier (Fig. 5.20) was an anatomist and palaeontologist who was one of the first to recognize patterns of extinction of species in the fossil record. To account for these he postulated the occurrence of catastrophic natural events such as widespread floods and episodes of rapid mountain-building. According to Cuvier, after such events new life forms would move in from other areas. Although he mentioned floods as examples of these catastrophes, he based his theories on field evidence alone, without recourse to any form of divine intervention. He was categorically not referring to Noah's Flood. Unfortunately for Cuvier this was not how many in the scientific establishment in Britain chose to interpret his work. William Buckland, who had taught Charles Lyell, believed that 'Geology is the efficient auxiliary and handmaid of religion' and he saw evidence everywhere of 'direct intervention by a divine creator'. Explicitly he equated the most recent of Cuvier's proposed inundations to the biblical Flood. This meant that catastrophism in Britain had religious implications that were not raised elsewhere in the world.

Lyell took advantage of the growing climate of opinion among scientists who were increasingly advocating a rational explanation for Earth's history to challenge the supporters of catastrophism. He branded them as being against rational argument and as adhering to the biblical chronology of the Earth as 5 kyr old and depending on a miraculous origin of Earth's phenomena. In many ways of course this was a gross misrepresentation of Cuvier who avoided all reference to religion in his scientific writing and who believed the Earth to be several million years old.

We can now see that there is no reason why an Earth many billions of years old should not be subjected to worldwide catastrophic events. The recently observed collision of Comet P/Shoemaker-Levy with the planet Jupiter and the increasing awareness of the possibility of a repeat of the Chicxulub meteorite impact is clear proof that paroxysmal events have a role in the history of the Solar System, including Earth. The sub-title of Lyell's book *Principles of Geology* was 'an attempt to explain the former changes of the Earth's surface by reference to causes now in operation' – hence the term uniformitarianism. Hutton's genius had been to recognize that changes in the Earth's surface that took place in the past could be compared with processes that were observable directly. In that sense the present *is* the key to the past. Lyell took this concept to the extreme and argued that all past events could be explained by the action of causes now in operation. His argument was flawed in that he did not allow for former causes to become redundant, or for new processes to arise, and there were no increases or decreases of rates with time. However, such was the eminence of Charles Lyell on the scientific landscape of the period that his views on uniformitarianism prevailed and were to exert a significant influence on geological thought for the next 150 years.

The idea of 'gradualism', that geological changes occur slowly over long periods of time, was the orthodoxy and those who advocated high-magnitude, low-frequency events were seemingly irrevocably associated with religious explanations for Earth processes. Today most geologists apply uniformitarianism in a different way. There is much evidence that the rates of various processes have not been constant throughout geological time. Geological thought therefore has moved to a position of what can perhaps be described as 'catastrophic uniformitarianism', which is a modification of uniformitarianism to take into account catastrophic (low-frequency/high-impact) events and their implications. This is evidenced by the now well-established occurrence of mass-extinction events throughout Earth history.

The aftermath of the uniformitarianism versus catastrophism controversy

So convincingly were the catastrophists discredited by Lyell and his followers that even into the middle of the 20th century there was reluctance among many geologists and geomorphologists to attribute Earth surface features to events that smacked of a catastrophe. Figure 5.21 shows Dry Falls, Washington State, USA, thought to have been the greatest waterfall ever to have existed on the planet. Some ten times bigger than the Niagara Falls, it forms a scalloped precipice nearly 6 km long with a drop over the rim of 120 m.

Around 18–20 kya during the North American glaciations the ice dam blocking the Clark Fork River in Idaho burst. The water behind the dam in glacial Lake Missoula, some 200 km to the east, drained rapidly, causing an outflow of cataclysmic proportions. This is now known as the Spokane Flood. Lake Missoula was comparable in size and volume to some of the present-day Great Lakes and probably emptied in a matter of days or weeks.

The maximum flow rate of the Spokane Flood is estimated by the US Geological Survey to have been about ten times the current combined flow of all the rivers of the world. Flow velocities in excess of 80 km per hour and a

Figure 5.21 Dry Falls, Washington State. © Laszlo Dobos/ Shutterstock.

water depth of 80 m above the rim of Dry Falls meant that the volume of water passing across the basalts of the Columbia Plateau carved out the area known as the 'Channeled Scablands'. This is a complex of rock-cut channels, cataracts, rock basins, gravel deposits and giant ripples – all products of the erosional and depositional effects of vast quantities of fast-flowing water. The term scabland is locally derived and refers to the areas of bare rock exposed by the erosion of the glacially derived soil known as loess. This leaves residual outcrops of the under-lying basalts as small 'scabs' on the landscape (Fig. 5.22).

A further effect of the flooding was giant ripple marks consisting of long gravel ridges, formed by the powerful currents flowing westwards from Lake Missoula. These co-exist with large-scale erratics, huge blocks of rock torn up by the floods and transported long distances before being dumped as the water velocity decreased.

The idea that the Scabland topography was caused by catastrophic flooding was first proposed by J Harlen Bretz in the 1920s but remained controversial for many years before general acceptance by the second half of the 20th century. Approval of the theory came gradually, starting first of all by the recognition of a plausible source of the huge amounts of floodwater that had to have been involved. In the Pleistocene period glacial Lake Missoula was estimated to hold approximately 2.66 cubic kilometres (km^3) of water in northern Idaho and western Montana and would have drained very rapidly south and south-west-ward over what is now the Channeled Scabland in Washington. An especially crucial piece of evidence for flooding on such a vast scale was the discovery of the giant current ripples and the realization of their significance.

Figure 5.22 Scabland topography, Washington State, USA.

The reason this model remained controversial for so long was because of the triumph of uniformitarianism over catastrophism a hundred years earlier. According to the orthodox but mistaken application of the uniformitarianism principle, cataclysmic processes such as the Spokane Flood were not considered suitable topics for scientific investigation. The eventual acceptance of this hypothesis has allowed a new understanding of the potential role of cataclysmic events in Earth and Solar System processes as evidenced by the role now recognized for impact cratering such as that at Chicxulub.

Lyell can be regarded as the short-term victor in the controversy between uniformitarianism and catastrophism. The vision he shared with Hutton of the primacy of time's cycle, the steady state theory of the origin and development of the Earth, prevailed as the orthodoxy for over a hundred years.

In the mid-19th century, following on from Lyell versus Cuvier, it was time for the next phase of new ideas to be expounded, for attempts to be made to measure geological time, both relative and absolute.

Chapter 6

The determination of relative time – beds in order

The successive series of stratified formations are piled on one another, almost like courses of masonry…

William Buckland (1784–1856), *Geology and Mineralogy Considered with Reference to Natural Theology,* **1836**

Time in geology can be considered in two distinct ways – relatively and absolutely. The relative age of a rock is its age compared to another – it can be older, younger or the same age as another rock. Essentially, relative time is a sequence of geological events. The absolute age of a rock is its age in years. In earlier chapters of this book when dealing with the rocks in north-west Scotland and those exposed in the walls of the Grand Canyon in Arizona these concepts were discussed. This chapter is concerned with the ways in which geologists determine relative time. Satisfactory methods of measuring the absolute age of geological processes had to await the development of accurate analytical techniques in the mid-20th century and will be dealt with in the next chapter.

The process of assessing the relative ages of rocks was part of the general recognition of the concept of geological time that developed from the 16th century onwards. This culminated in the establishment, by the end of the 19th century, of a generally accepted sequence of rocks in the Earth's crust, from oldest to youngest, known as the stratigraphical column. The Grand Canyon is perhaps the best physical illustration of that available to us. The steps whereby this stratigraphical column was built up is a fascinating example of the combination of a range of factors. These include field observations involving the identifications of lithologies, or rock types and their contained fossil species, along with philosophical musings on the nature of time.

Stratigraphy is the description, correlation and classification of sedimentary rocks. It is dependent on the law of superposition, the principle that in a sequence of undeformed sediments, a bed that overlies another bed is always the younger. This important principle was first formulated by Nicholas Steno

(1638–86), a Danish court physician living in Italy. His formulation followed on from an earlier observation that rock layers, or strata (singular stratum), were formed when particles in air or water fell to the bottom, leaving horizontal layers. Indeed the term sediment is from the Latin *sedimentum*, meaning settled. This is the principle of original horizontality, that sedimentary rocks form in the horizontal position and any deviations from this position are due to the rocks being disturbed later (Fig. 6.1a and b).

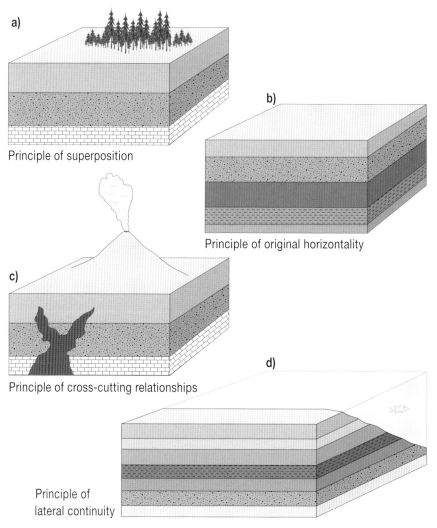

Figure 6.1 Nicolaus Steno's four principles of stratigraphy. **A)** The principle of superposition states that when sediments are deposited, those which are deposited first will be at the bottom, and so the lower sediments will be the older; **B)** the principle of original horizontality states that sediment is originally laid down flat; **C)** an event that cuts across existing rock strata is younger than the original strata; **D)** strata can be assumed to have continued laterally far from where they presently end. After Jones.

An innovative way of illustrating the principles of stratigraphy and the law of superposition is the study of the age and duration of old mining camps and towns like Bodie in the western USA from the latter part of the 19th century into the first half of the 20th century. Figure 6.2 shows the former gold mining town of Bodie in the Western Sierras of California, near the Nevada border. Today it is a 'ghost' town, a State Park, a National Historic site and a tourist attraction. It serves as a good example of the sort of settlement that sprang up all over the world as gold was mined from the middle of the 19th century in the new territories such as California, the Yukon, South Africa and Australia.

A study of the 'stratigraphy' of Brodie was carried out by Chas. B. Hunt of the US Geological Survey. Examining material left behind in the camp dump it was possible to establish when the settlement was established and how long it lasted for. Also, using information gleaned from the accumulated rubbish from the town waste dumps, Hunt was able to recognize four main stages of habitation. The oldest camps, those working before about 1900, are characterized by

Figure 6.2 The ghost gold mining town of Bodie in the Californian Sierras. © Hanze/ Shutterstock.

tin cans which have been soldered, that is, cans where the edges are sealed by a fusible metal alloy applied by a heated 'soldering iron'. The beer bottles in use at this time had necks which were finished by hand and designed for cork stoppers. A further characteristic of this period was square-headed nails, individually made by a blacksmith. From 1900 to the outbreak of the First World War in 1914, mining camps are characterized by round-headed nails, machine-made from wire. The beer bottles from this period are still hand-finished but by now are designed for metal caps rather than corks. Soldered tin cans continued to be used during this period.

In the 1920s and 1930s the style of tin cans and beer bottles changed. The tin cans were no longer soldered, instead the edges were crimped together, and the bottles all have machine-finished necks. In the period after the Second World War, those mining camps still operating were characterized by aluminium cooking utensils and beer cans rather than bottles.

Using the changes in bottles, cans and nails with time it is possible to erect a 'range chart' for these artefacts, showing their distribution over time. Such a chart is illustrated below (Fig. 6.3). In theory, having made a range chart in one location, this could then be used to date similar material from another location. In this example the cans, bottles and nails are being used in the same way as

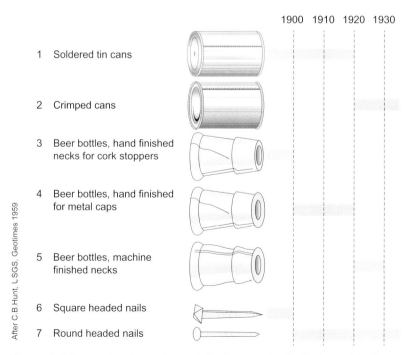

Figure 6.3 Range chart for artefact variation in types of nails, bottles and food cans used in Californian mining camps from the late 19th century to the 1930s.

fossil plants and animals are used to date and correlate sedimentary sequences. This is the principle behind the law of faunal succession, which states that fossil plants and animals succeed one another in a definite order.

The law of faunal succession

Fossils, the remains of ancient organisms found in some rocks, are an important tool used to establish the time sequence of a series of layers of sediments. The Greek philosopher Aristotle (384–322 BCE) recognized that fossil seashells he saw in rocks were similar to shells found on the beach indicating that the fossils were once living animals. Xenophanes of Colophon (560–478 BCE), a Greek scholar best known for his thoughts on religion and the significance of the Greek gods, was another who concluded from his observations that fossil shells embedded in rocks meant that water must once have covered all the Earth's surface and he suggested that the Earth had gone through alternating periods of extreme wetness and dryness. However, these early insights in classical times were lost to science with the decline of Greek civilization and it was not until the Middle Ages that their true significance was once again recognized and commented on. One of the first to do so post-classical times, perhaps surprisingly given his more obvious reputation as a painter and sculptor, was Leonardo da Vinci (1452–1519).

Leonardo da Vinci realized that marine organisms now found as fossils in the sediments of the Tuscan Hills in northern Italy were the remains of ancient life forms that had existed in the area when it was covered by the sea and they had simply been buried by mud on the seabed after their death. His insight into fossils extended to being among the first to realize that, as well as direct evidence, fossils could provide indirect evidence of their existence in the form of footprints, feeding trails and burrows. However, as with the observations of the early Greeks, none of his contemporaries seems to have developed the ideas he put forward and it was more than a hundred years after da Vinci before fossils and their significance were part of the scientific landscape again. Nicholas Steno played an important role in this rediscovery of the use of fossils. Steno was primarily a physician employed by Grand Duke Ferdinand of Tuscany in Italy and his principal research activities dealt with anatomy. It was in this capacity that he encountered the head of a giant shark caught by fishermen near the town of Livorno in 1666. The Duke ordered the head of the fish to be sent to Steno, who dissected it and published the results the following year, 1667, including the ferocious looking illustration reproduced here (Fig. 6.4).

Figure 6.4 Steno's drawing of a shark head and a so-called 'tongue stone'.

87

During his examination of the shark's teeth he noticed the similarities to objects known as 'glossopetrae' or tongue stones, found on the island of Malta. Steno proposed that the glossopetrae looked like shark teeth because they *were* shark teeth. These were derived from dead sharks whose teeth had been buried in the sand or mud of the seabed that was now dry land. While Steno's insight may seem to us a very obvious conclusion it was very much against the orthodoxy of the times and represented a major step forward in the appreciation of the significance of fossils in unravelling Earth history.

The remarkable English scientist Robert Hooke (1635–1703), arguably the greatest experimental scientist of the 17th century, was a contemporary of Steno. Hooke's contribution to the science of the late 17th century was immense, although now over-shadowed by his great rival, Isaac Newton. His interests covered physics, astronomy, chemistry, biology and geology and he was responsible for inventions such as the iris diaphragm, the universal joint and the balance spring in clock mechanisms. He is probably best summed up in his own words from one of his diary entries for 28 June 1680: 'spent most of time in considering all matters'.

Hooke was interested in fossils and compared them with living organisms and examined them with a microscope (Fig. 6.5) – the first person to do so. He

Figure 6.5 Robert Hooke's microscope.

noted that plant and animal material was capable of being altered by the action of mineral-rich water, essentially the same conclusion that Steno had come to in the case of the shark teeth.

Almost 200 years before Charles Darwin published his great work *On the Origin of Species*, Hooke had realized that fossils could be used to help illustrate the sequence of life on Earth from primitive forms to more complex organisms. He also recognized that certain fossils represented species that were no longer present on Earth and that other species may not have existed from the beginning of life on the planet:

> There have been many other Species of Creatures in former Ages, of which we can find none at present; and that 'tis not unlikely also but that there may be divers new kinds now, which have not been from the beginning (Robert Hooke, *Discourse of Earthquakes*, 1705).

There is a famous quotation attributed to Hooke's great rival Isaac Newton in a letter to Hooke: 'if I have seen further it is by standing on the shoulders of giants'. If taken at face value this is Newton acknowledging the contribution of others to his scientific endeavours. However, there is at least a possibility that it is a sarcastic reference to Hooke's shortness of stature. The two men were not on friendly terms. The point to be made here is that those geologists interpreting the stratigraphical and fossil record in the 19th and 20th centuries with such success were undoubtedly standing on the shoulders of giants such as Robert Hooke.

The making of the stratigraphical column

By the middle of the 18th century in continental Europe the first steps in the construction of a stratigraphical column were being taken. In 1756 Johann Gottlob Lehmann, a German geologist and mineralogist working in Saxony, proposed the existence of three distinct rock groups, based on the principle of superposition as advanced by Steno in the 17th century. The lowest group, the Primary, were essentially crystalline rocks, the intermediate or Secondary were layered rocks containing fossils, and the youngest sequence consisted of alluvial or river-derived unconsolidated sands and gravels.

A few years later in 1760 Giovanni Arduino, another mining specialist, this time working in Tuscany, in northern Italy, proposed a fourfold division of the succession. His oldest divisions, the Primary and Secondary, are broadly similar to those of Lehmann, but in addition Arduino proposed a further category, the Tertiary, to account for poorly consolidated, but still stratified and fossiliferous sediments underneath the unconsolidated alluvial sediments also recognized by Lehmann. By the end of the 18th century other workers in

Germany, Russia and Britain were using the concept of superposition when describing rock sequences and Steno's principles were being widely applied. The first stages in the construction of the stratigraphic column had been passed.

Steno's law of lateral continuity (*see* Fig. 6.1d) states that rock layers extend until they grade into those produced in adjacent environments. This lateral change is a facies, as discussed earlier in Chapter 4, and with the recognition of facies variability it was a relatively simple process to place large areas of Europe within a regional succession that was soon extended to parts of the rest of the world.

After the initial proposals of Lehmann and Arduino, the next stage in construction of a universal stratigraphic succession was by the German geologist and explorer Alexander von Humboldt (Fig. 6.6), famous for his observations of the cold current that flows north along the coast of Peru which now bears his name.

Humboldt was first to recognize the widespread occurrence of the 'Jura Kalkstein' or Jura Limestone in the Jura Mountains in eastern France (Fig.

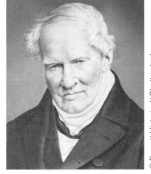

Figure 6.6 Alexander von Humboldt (1769–1859). Portrait as a young man by Friedrich Georg Weitsch 1806, in later years and commemorated by the Cuban Postal Service around 1969.

© Everett Historical/ Shutterstock

© Kiev.Victor/ Shutterstock

Figure 6.7 Limestone scenery in the Jura Mountains. © Fedor Selivanov/ Shutterstock.

6.7). Subsequently the rock type was recognized throughout Europe and the Jurassic System was established in 1795.

The faunal succession is the vertical arrangement of different fossils, corresponding to the series of sedimentary rock layers, the stratigraphic sequence, which contains the fossils. For convenience during mapping, rocks may be grouped into formations, a formation being a set of rocks that are horizontally continuous and share some distinctive feature of appearance and composition. As more and more were mapped in the early 19th century it became apparent that everywhere the faunal successions matched the sequences of sediments. In other words the fossils contained in a formation could be used to identify it, to 'fingerprint it' in effect. The fossils of different species appear in a definite order.

Increasingly it was observed that there was a consistent progression of fossils from primitive life forms to more advanced forms as the rock succession became younger. One of the first to exploit these observations for practical use was an English canal surveyor, William Smith (1769–1839), who from unlikely beginnings went on to make a unique contribution to stratigraphy and geological maps. Smith was born into modest circumstances as the son of a blacksmith and obtained only a rudimentary education. His father died when he was 8 years old and he was subsequently raised by an uncle. He began work as a surveyor and when working in coal mines in Somerset in the south of England he noted that the rock strata at the pit were arranged in a predictable pattern and the various strata could always be found in the same relative positions. He also noted that each particular bed could be identified by the fossils it contained. In the course of examining various natural outcroppings in his later work as a quarry and canal surveyor in the 1790s, Smith used not only the lithological characteristics of the rocks to fix their position in the succession, but increasingly their faunal or fossil content. As he extended his fossil collecting over a wider area he found that the same succession of fossil groups, from older to younger rocks, could be identified in many parts of the country and he used this information to correlate the successions at various localities he had studied.

Using his extensive knowledge of the rocks of the area and his large collection of fossils, he began to co-ordinate all the then known successions in England and Wales, culminating in 1815 with the publication of the first geological map of the whole of England and Wales and part of Scotland (Fig. 6.8). This constituted a remarkable achievement and the map provided the template for all other geological maps and the basis for geological surveys all over the world. With the completion of Smith's monumental map it was now possible to assume with a reasonable degree of certainty that correlation could be made between and among widely separated areas. The various rock types are denoted by different colours and the map bears a remarkable similarity to modern maps published for the region by the British Geological Survey.

It is possible to recognize the stratigraphic range of the main fossil groups from the Cambrian Period onwards (Fig. 6.9). There is considerable variation in the time intervals over which certain fossil groups are present in the geological record. Relatively advanced groups such as the fishes, reptiles and mammals (animals with a backbone, or vertebrates), appear at a much later stage than the less complex forms such as brachiopods and trilobites (invertebrates). The Phanerozoic Eon is considered to be the time of visible life starting at the base of the Cambrian, where organisms with hard parts appeared in abundance for the first time – the so-called Cambrian explosion. The Phanerozoic as we will see extends for over 500 myr, but the preceding eon, the Precambrian existed for 4 byr and the unravelling of this vast expanse of time is one of the great successes of the Earth sciences – but that is an epic for the next chapter.

Thus with all the fundamental concepts in place by the beginning of the 19th century, rock successions in different locations could be equated, not just in parts of Britain as demonstrated by William Smith, but globally. Before the end of the 19th century there would be a stratigraphic column that allowed geologists working anywhere on the globe to correlate their stratigraphy with other workers elsewhere.

In England the Coal Measures were coal-bearing strata that had been exploited for centuries and by the mid-19th century the beds were being

Figure 6.8 First geological map of Great Britain, published by William Smith, 1815.

extensively mined to fuel Britain's Industrial Revolution. Two English geologists, William D. Conybeare and William Phillips, established the term Carboniferous (carbon-bearing) in 1822 to reflect the extensive coal deposits in the system. In the same year the Belgian geologist Jean d'Omalius d'Halloy used strata exposed in the Paris Basin to designate the Cretaceous

PERIOD		ANIMALS	PLANTS
Quaternary			
Tertiary			
Cretaceous			
Jurassic			
Triassic			
Permian			
Carboniferous	Pennsylvanian		
	Mississippian		
Devonian			
Silurian			
Ordovician			
Cambrian			

Figure 6.9 Stratigraphic ranges and origins of some major animal and plant groups. After USGS.

Figure 6.10 White Cliffs of Dover, Cretaceous Chalk. © MrPics/ Shutterstock

System. He based the system on the predominant white limestones or chalk as seen in the White Cliffs of Dover on the south coast of England (Fig. 6.10) and extensively in France, Belgium other coastal regions around the North Sea and the Baltic. The term Cretaceous is from the Latin *creta*, meaning chalk.

The Triassic, originally the Trias, was named by the German palaeontologist Friedrich August von Alberti in 1834. The system was derived from the Secondary category of Lehmann, who was mentioned earlier as the first person to propose a sequence of rock types for a region. The rock types recognized in his subdivision were subsequently redivided into three distinct lithostratigraphic units – the Bunter Sandstone, the Muschelkalk Limestone and the Keuper Marls and Clays – hence the Trias, meaning threefold. Rocks of this age are commonly characterized as desert sediments and saw the first appearances of the great reptiles in the fossil record.

The British geologist Sir Roderick Impey Murchison was a remarkable man who by 1840 had made a substantial contribution to stratigraphy and along with Adam Sedgwick had already designated several systems of the stratigraphical column. These two men who were close collaborators in the early phases of their careers were to become bitter antagonists in one of the great geological disputes of the 19th century. In 1841 Murchison was in Russia on an expedition sponsored by Tsar Nicholas 1 to attempt to explain the two successions of 'red beds' that lay above and below the rocks of Carboniferous age. He recognized post-Carboniferous red beds overlying the classic Carboniferous limestones of the Moscow Basin and he decided these rocks were part of a new system, the Permian, named after the region of Perm and bridging the gap between the Carboniferous and Trias seen in western Europe.

As with the later Triassic, the Permian rocks are characteristically desert sediments. The whole period was dominated by the supercontinent Pangaea which had vast desert areas in its continental interior. During the Permian the reptiles rose to dominance, better able to cope with the dry conditions, and the period ended with the greatest mass extinction in Earth's history.

Around 90% of marine species and 70% of terrestrial species died in this event. While there were probably a number of causes for an extinction episode on this scale, there is evidence that the prolonged eruptions of flood basalts over many thousands of years during this period, forming the Siberian Traps, would have been a major source of environmental stress. Additional factors were the extreme aridity and the reduction in coastal environments associated with the amalgamation of the continents into the supercontinent Pangaea.

This event marked the end of the Permian and the beginning of the Triassic and it would be well into the latter period before the Earth recovered its biodiversity. The extinction was a watershed in life on Earth, particularly for marine life. The extinction event at the end of the Cretaceous, sometimes referred to as the K–T event, since it occurred at the boundary between the Cretaceous (K) and Tertiary (T) periods is perhaps better known to the general public. In fact the K–T event, which accounted for the demise of the dinosaurs, was relatively insignificant in comparison to the Permian extinction.

The cryptic rocks of Wales

If, as we have seen, the early part of the 19th century was dominated by the clash of Lyell and Cuvier over the processes of uniformitarianism and catastrophism, then the equivalent intellectual clash of the second half of the century was the long-standing and often bitter dispute that developed between Sedgwick and Murchison concerning the naming of the older rocks of the stratigraphical column below the Carboniferous Period. By around 1840 significant progress had been made in the building of a stratigraphical system which was of use across the globe. The system immediately below the Carboniferous, the Devonian, had been named by Murchison and Sedgwick, jointly, in 1840, with the Silurian below that already designated by Murchison in 1835 and named after the Silures, a tribe in ancient Wales.

Also in 1835 Sedgwick had proposed the name Cambrian for a sequence of poorly fossiliferous rocks lying immediately above the so-called Primary rocks in north Wales. Cambria was the Roman name for Wales. It was the sequence of rocks from these designated Cambrian through the proposed Silurian and ending in the red beds of the Devonian which were to prove controversial in the following nearly 40 years. It was to be 1879 before a compromise was finally worked out, by which time both protagonists were dead after a long and at times acrimonious feud.

Adam Sedgwick was born in 1785, the third of seven children of an Anglican vicar, and after graduating from Cambridge in mathematics and theology he took holy orders in 1817 and rather bizarrely became Woodwardian Professor of Geology at Cambridge in 1818, despite having no formal training in the subject. He held the chair until his death in 1873. He is alleged to have remarked, on his appointment to the post, 'Hitherto I have never turned a stone; henceforth I will leave no stone unturned'. Despite his late introduction to the subject, Sedgwick quickly became an active researcher in stratigraphy and palaeontology, carrying out fieldwork all over Britain while at the same time vastly enlarging the geological collections of the university.

One of his important collaborators in the early stages of his career was Roderick Impey Murchison, described at times as a 'gentleman geologist'. Murchison was born in Ross-shire, Scotland, in 1792 and came from a wealthy family. His father having died when Murchison was 4 years old, the young Roderick was sent to military college to be trained for a career in the army. This subsequent career involved service under the Duke of Wellington in the Peninsular War, including the epic retreat to Corunna and the final battle there under Sir John Moore (1809). Following his period of military service (1808–16) Murchison married and settled eventually in Co. Durham. Here, after becoming acquainted with Sir Humphry Davy, the famous chemist and inventor of the Davy lamp for use in coal mines, he became interested in geology and joined the recently formed Geological Society of London. He became an active member and met fellow pioneers of geology such as Adam Sedgwick, William Conybeare, William Buckland and Charles Lyell. These years are sometimes referred to as the 'heroic age of geology', a reference to the pioneering work being done in establishing the main boundaries of the stratigraphical column.

Throughout the 1830s both Sedgwick and Murchison had been investigating the complex succession of rocks found in Wales, rocks made even more difficult to unravel because of their extensive folding and faulting. In 1831 they began working on the sequence of beds beneath the Old Red Sandstone, rocks which had been included in the Carboniferous by Conybeare and Phillips some 9 years previously. In 1835, after what had started as a collaborative effort to solve a difficult geological conundrum, they presented two distinct subdivisions of the pre-Carboniferous rocks in Wales:

- In north Wales, working stratigraphically upwards from the base of the post-Primary succession of poorly fossiliferous sediments, Sedgwick identified a sequence defined principally by its various lithologies, which he designated Cambrian.
- In south Wales, working stratigraphically downwards in the more fossiliferous pre-Old Red Sandstone rocks, Murchison was able to identify a succession of strata containing a well-preserved fossil fauna, including trilobites and brachiopods, which he called Silurian.

In a relatively short time this system was expanded as more and more localities containing the characteristic fauna were identified throughout Europe. It soon became evident that the 'Silurian' in northern Wales was coincident with many of the strata of Sedgwick's 'Cambrian'. With the Cambrian based mainly on the properties of the rocks, the presence of Silurian fauna was creating difficulties with correlation. Murchison eventually claimed that the entirety of Sedgwick's Cambrian was really part of the Silurian, rather than a separate system. The resulting quarrel meant the two men were permanently estranged and the issue was never resolved in their lifetime. The compromise solution was put forward by Charles Lapworth (Fig. 6.11), an English geologist, who with careful use of fossils, particularly graptolites, proposed in 1879 the designation Ordovician for the sequence of rocks representing the upper part of Sedgwick's Cambrian and the lower part of Murchison's Silurian.

This fossil evidence from Wales and also from Scotland finally enabled the separate systems Cambrian, Ordovician and Silurian to be confirmed. The table below shows the position of the Ordovician Period relative to the original Cambrian and Silurian Periods.

Silurian (Murchison)	Silurian
	Ordovician (Lapworth)
Cambrian (Sedgwick)	Cambrian

Figure 6.11 Plaque on the wall at Birkhill cottage, dedicated to Charles Lapworth.

Lapworth was in many ways a remarkable man. Although largely self-taught in geology and doing much of his important early research when working as a school teacher, he bequeathed a substantial legacy to British geology. In designating the Ordovician System he not only resolved one of the bitterest geological controversies of the 19th century, he also made a significant contribution to the worldwide stratigraphical column.

As with many scientific disputes over the centuries, not just in geology, with the passage of time one wonders whether it was as important as it seemed to the antagonists at the time. Geology moved on without them, with the Ordovician becoming an established part of the stratigraphical column by the early years of the 20th century.

Completion of the Phanerozoic timescale

By the 1850s the development of the relative timescale was almost complete and it was beginning to be recognized that there were a number of important geological boundaries, defined on fossil content that were common and recurrent across the world. The principle of faunal succession was proving to be consistent and thus any major changes in faunal characteristics were viewed as significant. Based on proposals by Sedgwick and others in 1838, the term Palaeozoic Era was used to cover the periods from Cambrian through to

Permian. These were sediments containing recognizably primitive life forms. Palaeozoic is derived from the Greek meaning 'old life forms'. The Palaeozoic is that time between the Cambrian 'explosion of life' and the great Permian extinction. Following this in 1840, William Phillips, who with William Conybeare in 1822 had named the Carboniferous System, proposed the term Mesozoic Era and Cenozoic Eras meaning 'middle life forms' and 'new life forms' respectively. The Mesozoic Era is bracketed by the Permian extinction at its base and the K–T extinction event at the end of the Cretaceous. These three eras were then collectively referred to as the 'Phanerozoic Eon', originally referring to the period of visible life, but now generally taken as the time of complex life; that is from just before the Cambrian to Recent. The time preceding the proliferation of complex life on Earth is referred to as the 'Proterozoic Eon', meaning earlier life.

Sir Charles Lyell and the subdivision of the Tertiary

Sir Charles Lyell was mentioned in an earlier chapter as a leading proponent of the role of uniformitarianism in Earth history, refusing to consider any role for what was termed catastrophism as put forward by the French scientist Georges Cuvier. While the contemporary view of Lyell is probably unduly negatively influenced by his refusal to countenance any role for catastrophic events in Earth history, he nevertheless played a major role as a stratigrapher in the subdivision of the Tertiary in the first half of the 19th century.

In the late 1820s Lyell travelled with Murchison in the south of France and Italy and in these areas concluded that the rock layers or strata could be classified (dated) on the basis of the proportion of marine shells they contained. Further work in the London Basin and Paris Basin showed that Giovanni Arduino's proposed Tertiary System (put forward in 1760), based on the Tuscan Hills of Italy could be applied to all these localities. Accordingly Lyell proposed subdividing the Tertiary into three parts, the Pliocene, Miocene and Eocene. These names were derived from the Greek descriptions of the proportion of present-day species in the fauna. *Eos* means dawn, hence the Eocene which means the beginning of modern life forms, while *meion* means less and *pleios* means more, hence the Miocene and Pliocene. Lyell had recognized that in the Tertiary modern species appear as fossils, becoming more abundant in younger sediments. For example, 3% of Eocene species are alive today, while as many as 30–50% of Pliocene species exist today. Progressively older rocks yielded fewer and fewer forms with living counterparts. In 1839 Lyell proposed the term Pleistocene from the Greek *pleiston* meaning most for the youngest rocks of the Tertiary, those with the most species still present in modern environments.

The youngest materials, often unconsolidated and described by earlier workers such as Arduino and Lehmann as 'alluvium' were now considered as deserving a place in the formalized stratigraphical column which was in the course of construction. In 1829 therefore the term Quaternary was proposed by Jules Desnoyers of France to cover the post-Tertiary rocks of the Seine Basin. It now comprises the glacial period, the Pleistocene, and the youngest epoch, the one we are currently living through, the Holocene.

The subdivision of the lower part of the Tertiary was completed in 1854 and 1874 when two German scientists Heinrich Ernst Beyrich and Wilhelm Philipp Schimper proposed the terms Oligocene and Palaeocene respectively, from the Greek *oligos* meaning few and the Greek *palaios* meaning old. Figure 6.12 shows the end results of the subdivision of the Tertiary.

ERA	PERIOD	EPOCH	Start date (mya)
CENOZOIC	Quaternary	Holocene	(0.01)
		Pleistocene	(1.6)
	Tertiary	Pliocene	(5)
		Miocene	(23)
		Oligocene	(35)
		Eocene	(56)
		Palaeocene	(65)

Figure 6.12 Subdivisions of the Tertiary. In modern stratigraphic terminology Tertiary has been replaced by a lower division, the Palaeogene consisting of the Palaeocene, Eocene and Oligocene, and an upper division the Neogene comprising the Miocene and Pliocene.

The stratigraphic column

The stratigraphic column shown (Fig. 6.13) is a modern generalized version that could be applied anywhere in the world and it shows the main subdivisions and periods of geological time as recognized by geologists. The periods are in order from oldest at the bottom to youngest at the top, showing their relative age. The methods used to calculate the ranges in years of the various subdivisions, their absolute age, will be the subject of the next chapter. The stratigraphical column is the product of the observations of many geologists, mining engineers and surveyors working with rocks in a variety of capacities. They examined sequences of sediments and fossils and over time recognized patterns and made correlations from one region to another. Investigators working across the world had realized that a number of major geological boundaries, defined on the basis of their contained fossils, were common and recurring, irrespective of their geographical location.

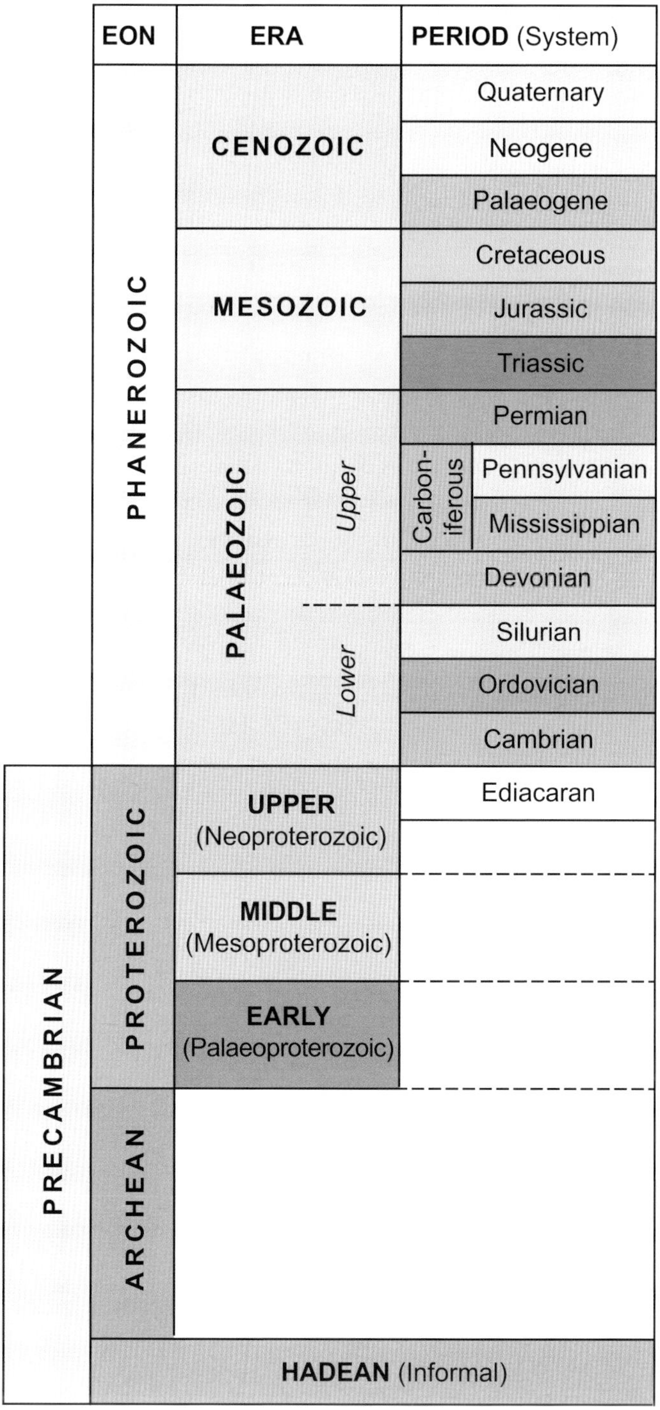

EON	ERA	PERIOD (System)		
PHANEROZOIC	**CENOZOIC**		Quaternary	
			Neogene	
			Palaeogene	
	MESOZOIC		Cretaceous	
			Jurassic	
			Triassic	
	PALAEOZOIC	*Upper*		Permian
			Carboniferous	Pennsylvanian
				Mississippian
				Devonian
		Lower		Silurian
				Ordovician
				Cambrian
PRECAMBRIAN	**PROTEROZOIC**	**UPPER** (Neoproterozoic)	Ediacaran	
		MIDDLE (Mesoproterozoic)		
		EARLY (Palaeoproterozoic)		
	ARCHEAN			
	HADEAN (Informal)			

Figure 6.13 Geological column showing the main subdivisions and periods of geological time. After Wyse Jackson

Louis Agassiz and worldwide glaciations

The final decades of the 18th century and the first half of the 19th century were momentous for the young science of geology. There was advancement on so many fronts that it must have been difficult for those involved in the research to keep up to date with all the developments. As well as the usual cut and thrust of scientific debate, a process that still takes place in contemporary science, there was, however, at that time a pervasive undercurrent of theology which influenced how research was conducted and results published. In Europe the influence of the Christian churches had an important bearing on how science progressed. It would be wrong, however, to assume that the influence of the church (of whatever denomination) was always so heavy-handed. Many of those involved in the great scientific advances of the 18th and 19th centuries had sincerely held views on topics such as the Creation and the creation of species. As a result of this some had difficulties reconciling ideas such as the immensity of geological time or the evolution of species with their religious beliefs. One aspect of the biblical account of the early days of the Earth that still had significant adherence was the literal truth of Noah's Flood as a worldwide deluge. This flood was held responsible for both the movement of impossibly large blocks long distances and also for the deposition of extensive layers of various muds and sands. After the demise of Neptunism and Creationism, it seemed that Diluvialism, the belief that the Earth had been shaped by a universal flood, had still to be addressed.

By the 1840s this problem was about to be addressed by a young Swiss scientist Louis Agassiz. Agassiz (Fig. 6.14) was born on 28 May 1807, the son of a minister, in the village of Môtier, in the French-speaking part of

Figure 6.14 Louis Agassiz (1807–73).

Switzerland. Agassiz was educated as a physician, like many naturalists of the time. After receiving his medical degree from the University of Erlangen in 1830, he went to Paris on 16 December 1831 to study comparative anatomy under Georges Cuvier. Agassiz's association with Cuvier was short-lived as Cuvier died in the Paris cholera epidemic of 1832. However, he had been so impressed with Agassiz's work on fossil fishes that he turned over to Agassiz his own notes and drawings for a planned work on this topic. With the publication in 1834 of the first part of this vast work on the fossil record of fishes, *Poissons fossiles* (five volumes published over a period of 11 years), Agassiz's reputation began to grow in the scientific community. He took up a professorship at the Lyceum of Neuchatel in Switzerland, where for 13 years he worked on many projects in palaeontology and glaciology.

Agassiz took up the study of glaciers in 1836 and was guided by colleagues Ignatz Venetz and Jean de Charpentier to examine the geological features of his native Switzerland. As early as 1821 Venetz had suggested that the Alpine glaciers are the remnants of former much more extensive bodies of ice which had left the ridges and hummocks of boulders and sand far from the terminations of present glaciers. Agassiz noticed the features left by glaciers on the landscape: wide U-shaped valleys (Fig. 6.15A), scratches or striations and the polishing of rock surfaces, and mounds of debris called moraines (Fig. 6.15B), formed by the power of the ice. He

Figure 6.15A U-shaped Yosemite Valley, California. © oooperman/ Shutterstook

Figure 6.15B Glacier showing surface moraine material. © Zbynek Burival/ Shutterstock

realized that in many places these signs of glaciation could be seen where there were no glaciers in the area, large glacial erratic boulders carried long distances (Fig. 6.16A), moraine deposits left after the glacier responsible for their formation has retreated (Fig. 6.16B), while Figure 6.16C shows sands and gravels deposited in a glacial meltwater lake.

Features such as moraines and erratic boulders had been explained previously as resulting from flooding, often the biblical Flood, and were frequently collectively referred to as diluvium.

Agassiz became a powerful advocate of the theory that a great ice age had once gripped the Earth, and published his ideas in *Études sur les glaciers* in 1840. In the book he describes how snow changes to glacial ice and describes in detail the formation of moraines along the length of the glacier and at its termination. Following on from his descriptions of contemporary glaciations he lists the criteria for identification of previously glaciated areas, including the recognition of former lateral and terminal moraines, erratic blocks and striated and smoothed rock surfaces. His later book, *Système glaciare* (1847), presented further evidence for this theory, gathered from all over Europe. Agassiz later found even more evidence of glaciation in North America. He therefore did not envisage the extension of the present glaciers during an ice age but instead saw them as remnants of an ice age.

Figure 6.16A Twistleton Scar glacial erratic, Yorkshire Dales National Park. Here a sandstone boulder is perched on the limestone pavement. © Phil MacD Photography/ Shutterstock

Figure 6.16B Lateral moraines above Lake Louise, Alberta, Canada. Photograph by Wilson 44691.

Figure 6.16C Sands and gravels deposited in a glacial meltwater lake, Ireland. The height of the exposed face is approximately 4m. Photograph by J.S.Shanks.

The immensity of geological time

By the last decades of the 19th century the stratigraphical column was essentially complete and applicable across the world. A feeling for the immensity of geological time was by now the orthodox scientific view. There had been a general move away from a strictly literal interpretation of the Bible when dealing with matters pertaining to the formation of the Earth and the development and evolution of species. It would be a mistake, however, to assume that the first half of the 19th century was simply a procession of transformative discoveries leading relentlessly to scientific progress. Charles Lyell for example did not believe in evolution and Adam Sedgwick's views on the subject were strongly bound up with his views on a divine creation. Both these men had difficulty dealing with the ideas of species development inherent in Darwin's theory. Darwin was using the larger scope of geological time that came from Lyell's ideas to put forward his plan for the evolution of species, but Lyell could not agree with the concept of mankind being descended from apes. So insight in one aspect of Earth history did not necessarily lead to the next stage in the understanding of geological time and all its implications.

Having successfully placed the rocks in their relative order in time, the emphasis now moved towards gauging the extent of geological periods in years, the absolute age of the Earth.

A quotation from the writer Franz Kafka, '*L'éterniteé, c'est long...surtout vers la fin*', which translates as 'eternity is a long time, especially towards the end', can be reversed from a geological perspective. Kafka's comment about eternity being long, especially towards the end, could be applied equally to the beginning. What was becoming increasingly obvious as progress was made in constructing the Phanerozoic or visible life timescale, was the extent of time involved in the era before visible life, the Precambrian. The unravelling of the geological history of these poorly fossiliferous rocks was now the challenge facing the science, and following the so-called 'heroic age of geology' involving the likes of Sedgwick, Murchison and Lyell it was time for a new generation of champions to take on that challenge.

Chapter 7

Measurement of absolute time – the age of the Earth

The great workman of nature is time…

Georges-Louis Leclerc, Comte de Buffon (1707–88) *Les Animaux Sauvage,* **1756**

The beginning of geological time

Figure 7.1 is an image sent back from the Hubble Telescope that was carried into space by a Space Shuttle in 1990 and remains in operation. Since the orbit of the telescope is outside the distortion of the Earth's atmosphere it is capable of taking high-resolution images with almost no background light. The result has been some of the most detailed visible-light images ever taken, thus allowing a view into deep space and therefore into deep time. Among other major advances made possible by the Hubble Telescope is an accurate determination of the rate of expansion of the universe.

The Great Nebula in Orion, about 1,500 light years away within our own galaxy the Milky Way, contains newly discovered regions where stars are being created – so-called stellar nurseries – and Figure 7.1 shows a number of young stars. Some of the stars are surrounded by gas and dust trapped as the stars formed, but left in orbit around the star. These are possibly proto-planetary discs or proplyds that might evolve into planets.

The evidence provided by images like this of the likely processes of star formation in the galaxy is important in understanding the possible stages of the formation and development of our Sun and Solar System, which obviously includes the formation of planet Earth. As already noted in Chapter 1, looking out into space means also looking back in time, and it is to space we must turn for insight into the earliest days of Earth. A protoplanetary disc is a disc of dense gas which surrounds a newly formed star and typically forms from molecular clouds consisting primarily of molecular hydrogen. When parts of this cloud reach a critical size, mass or density they begin to

Figure 7.1 The Great Nebula in Orion. The star-forming region's glowing gas clouds and hot, young stars are on the right of the photo. The Orion Nebula is some 1,500 light years away. NASA image.

collapse under their own weight and this collapsing cloud is called a solar nebula. This process is thought to have been the source of our Sun with its associated planets, including Earth. The cloud includes within it dust grains that condensed from the gas phase in the cooling outflows of stars from events such as red giants and supernovas. A red giant is a late stage in stellar evolution when as the star exhausts its resources of hydrogen fuel it begins to expand. In the very late stages the outer layers of the star are blown off in huge clouds of gas and dust. A supernova is a stellar explosion, extremely luminous and producing a burst of radiation that may briefly outshine an entire galaxy. A supernova may be triggered by the gravitational collapse of a massive star. The explosion expels much or all of the star's material at velocities up to 30,000 km per second and generates a shock wave which sweeps up an expanding shell of gas and dust called a supernova remnant. Both red giants and supernova are therefore likely to have contributed to the molecular cloud from which the Sun and planets were derived.

While most of the dust cloud in the solar nebula was vaporized as the growing Solar System heated up, a small portion of it was preserved, usually protected inside asteroids. Occasionally pieces of these asteroids fall to the Earth and are called 'primitive meteorites'. Within these meteorites are tiny

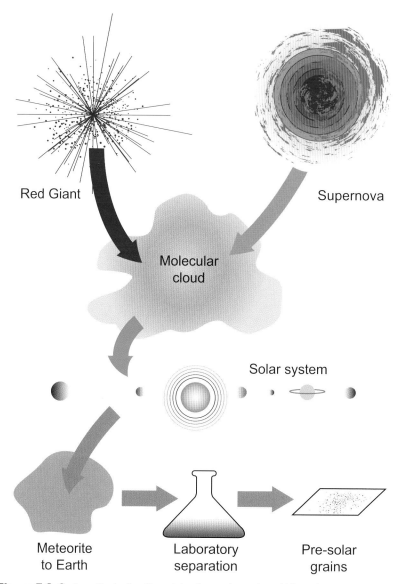

Figure 7.2 Cartoon illustrating the origin of presolar grains within a planetary nebula, After Manavi Jodhar'

dust particles which are examples of presolar stardust, presolar referring to their origin prior to the existence of the Sun. Figure 7.2 illustrates the entire process.

These grains in meteorites formed within the outflowing and cooling gas and dust clouds from stars which evolved earlier than the Sun. They are known to be presolar because they have exotic properties, principally abnormal values of certain isotopes. An isotope is a version of an element which is heavier or lighter than normal due to differing numbers of particles called neutrons, and the discovery of the existence of isotopes at the end of the 19th century had important implications for the measurement of absolute time and the calculation of the age of the Earth. These isotopic differences are suggestive of the processes of star genesis and the grains are contained in meteorites that formed early in Solar System history. They consist of minerals described as refractory, which means they are resistant to heat and chemical change. They have survived the collapse of their parent star, the subsequent collapse of the molecular cloud that formed the Solar System and eventually atmospheric entry when the asteroid or meteorite fell to Earth. Given such a turbulent history they have valuable insights to offer into the formation of the early Earth and the start of geological time.

Stardust in meteorites

Stardust grains are characteristically very small. The largest grains can be as big as a few tens of micrometres (μm) in diameter (a micrometre is one millionth of a metre or a thousandth of a millimetre) but these are very rare and most grains are less than a micrometre in size. There are a variety of types of stardust grains with distinct chemical compositions, though the most important are diamonds and silicon carbide:

- **Diamond:** Diamonds are the smallest presolar grains that have been identified; they are typically about 2 nanometres (2 billionths of a metre) in diameter and only contain around one thousand atoms. They are believed to be presolar because they contain isotopically unusual xenon and nitrogen. Xenon is one of the so-called inert gases which also occurs in trace amounts in the Earth's atmosphere. The composition of the xenon suggests that the grains originated in supernova explosions. However, only about one diamond in a million contains any xenon atoms at all, so it is possible that most of the diamonds found in stardust grains formed somewhere else.

- **Silicon carbide:** Presolar silicon carbide (SiC) is probably the best-studied of any of the known presolar grain types. SiC grains range in size from about 0.1 μm up to as large as 20 μm. Just about every element ever measured in presolar SiC grains has been found to be isotopically unusual, including silicon, carbon, nitrogen, magnesium, calcium, titanium and many others. Most presolar SiC is believed to have formed in 'AGB' stars, a certain type of very old red giant star that is rich in carbon, but some of the grains have isotopic signatures indicating other stellar sources, including supernovae and novae.

As well as diamond and silicon carbide, presolar grains include graphite, various oxides (principally aluminium oxide which is the mineral corundum) and silicates (minerals such as olivine and pyroxene commonly found in terrestrial rocks such as basalts).

Since these presolar grains are the oldest materials in the Solar System and formed before the formation of the Sun, they provide a window into the processes of star formation and the composition of the Sun's parent molecular cloud. The ages of the meteorites they are found in give an indication of the time of the birth of the Solar System and from the point of view of this study a starting point for geological time and the age of the Earth.

The search for the age of the Earth

By the last decades of the 19th century the succession of rocks worldwide was well established and a stratigraphical column had been developed that placed the principal periods of Earth history in order and also showed the gradual evolution of the main life forms as indicated by the fossil remains present in the rocks. Attention in geology now switched to the measurement of the absolute age of the rocks of the geological column, that is their age in years before the present day. Attempts at calibrating the stratigraphical column were part of a wider desire to know the age of the Earth, something that had exercised enquiring minds for hundreds of years.

Determination of the age of the Earth: accumulation processes

The basis of uniformitarianism is that the processes operating today are essentially the processes that have always applied. This means that the rates of various ancient processes could be considered to be the same as those operating today and so it should therefore be possible to calculate the age of the Earth on the basis of the total accumulation accrued as a result of this known rate. The first recorded instance of this technique being applied has already been mentioned – the estimate by Herodotus in the 5th century BCE of tens of thousands of years as the time taken for the sands and silts of the Nile Delta to accumulate. Since Herodotus there have been many such attempts to estimate the age of the Earth using various means of analysis such as the rate of accumulation of salt in the oceans or the calculation of the total thickness of sediment in the geological record divided by the average rate of deposition. All have had flawed arguments to a greater or lesser degree, usually because of an incomplete understanding of the totality of the process being considered.

In 1715, the English astronomer Edmond Halley put forward a plan to estimate the age of the Earth by calculating the rate of salt accumulation in the sea and thus working out the age of the ocean from its current salinity. Halley reasoned that since rivers continually wash small amounts of dissolved salts into the oceans, then progressively the oceans will become saltier. He further reasoned that this process could not have taken place over a relatively short period of a few thousand years (as required by the biblical timetable proposed by Archbishop Ussher) otherwise the oceans would still be mostly fresh water. Conversely the Earth could not be of great antiquity or the oceans would be saturated with salt like the Dead Sea.

In the event Halley never actually used his method to compute an age of the Earth. In 1899 an Irishman, John Joly, returned to Halley's plan and made another attempt to gauge the age of the Earth by estimating the rate of salt accumulation. John Joly was Professor of Geology at Trinity College, Dublin, from 1897 until his death in 1933. During his lifetime he engaged in a wide range of scientific projects, but it is his interest in the calculation of the age of the Earth that we are concerned with here.

Joly, assuming the ocean was a closed system with no salt being lost once it had accumulated, calculated that the Earth had formed and the oceans developed some 80–90 mya. This was an underestimate since it does not take into account the various ways that salts are naturally removed from the oceans. This subsequent realization that the ocean is not a closed system and that there is a continual loss of salt to sedimentary processes undermines the whole basis of the method.

Equally imaginative, but also flawed, were the attempts made to estimate the age of the Earth by calculating the total thickness of the sedimentary record and dividing it by the average rate of accumulation of sediment. One notable attempt using this approach was by John Phillips in 1860. Phillips was an English geologist and the nephew of William Smith. In 1840 he had proposed the terms Mesozoic and Cenozoic for the middle and new life form eras of the stratigraphical column. Using rates of erosion and deposition gauged from the Ganges delta and with a total figure for the total thicknesses of the fossiliferous systems of the geological column of 72,000 feet (about 22,000 m) he calculated the age of the Earth at approximately 96 myr. Like all attempts at using the rate of sediment accumulation as an 'hourglass' to measure geological time it did not take fully into account the different accumulation rates associated with different sedimentary environments. Nor did it recognize the significance of the many breaks in the stratigraphical record. While this figure therefore turned out to be a serious underestimation of the age of the Earth, it was an important contribution to a debate about the extent of geological time. By the 1840s this controversy had begun in earnest and was to continue for the remaining decades of the century and beyond.

The cooling Earth

If one person could be described as a scientific colossus of the late 19th century, it would surely be the imposing figure of William Thomson, 1st Baron Kelvin of Largs to give him his full title (Fig. 7.3). He was reputed to have more letters after his name than anyone else in the country and his influence on the development of physics during his lifetime was immense.

Kelvin was born in Belfast where his father taught mathematics at a local school before moving his family to Glasgow where he took up the Chair of Mathematics at the University. The young William Thomson, fourth child in a family of seven, was a precocious student of mathematics who matriculated into Glasgow University at the age of 10 in 1834. He published his first scientific article at the age of 16, a defence of 'The analytical theory of heat' by the French scientist Jean-Baptiste-Joseph Fourier. This was a prelude to a lifelong interest in the physics of heat which led him eventually to his involvement in the controversies surrounding the calculation of the age of the Earth. He was a co-author of the Second law of Thermodynamics, recognizing that the

Figure 7.3 William Thomson, Baron Kelvin of Largs (1824–1907).

conversion of one form of energy into another is never perfectly efficient. Some energy is always lost as heat, meaning that every system is constantly running down and will eventually stop.

The unit of temperature on the absolute temperature scale is the Kelvin and his mathematical analysis of electricity and magnetism was a major factor in his inventions used in the cables for submarine telegraphy in the 1850s. The first successful trans-Atlantic cable was laid in 1866 from Valentia Island, off County Kerry in Ireland, to Newfoundland. Kelvin's rewards for this achievement were recognition by a wider public, a knighthood from Queen Victoria herself and (as a director of the Atlantic Telegraph Company and a well-paid consultant for similar projects elsewhere in the world) substantial financial gain. Applying his thermodynamic principles to the problems of heat flow Kelvin attempted in 1862 to calculate the age of the Earth. In this he was following on from the ideas of the Compte de Buffon in the middle of the 18th century (see Chapter 4). His assumption was that the Earth and Sun were originally equally hot and that both had been cooling ever since. He also assumed that the Earth was incapable of producing new heat. He calculated that the Earth would take between 20 and 400 myr to cool to its present state, with 98 myr being the likely age. In 1868 he revised these figures down to an age of not more than 100 myr and in 1899 he further reduced the figure to 20–40 myr with the probability of being nearer to 20 than 40. Kelvin's assumption that no new heat was involved in the cooling history of the Earth was challenged by the Curies' discovery in 1903 that radioactive decay gave off heat. As a result of this and other findings it was apparent that naturally occurring radioactive elements such as uranium in common rock-forming minerals in the Earth's crust produce heat continuously. There is more heat in the Earth than Kelvin was aware of and his assumed rate of cooling was therefore fundamentally flawed.

The discovery of radiation

In 1895, Wilhelm Conrad Röntgen (Fig. 7.4), a German physicist working with an early type of cathode ray tube (the tubes found in older types of television sets), noticed that a specimen of a barium compound nearby in the laboratory began to glow when the current was turned on in the tube. He realized the tube was producing a previously unknown form of energy that was capable of passing through otherwise opaque materials. Since he had no idea of the provenance of this form of energy he named them X-rays.

Following this discovery of X-rays by Röntgen a train of events began that were to revolutionize physics and medicine, and also to extend the measurement of geological time to thousands of millions of years, a period undreamed of in all the previous speculation about the extent of Earth history. Röntgen had

Figure 7.4 Wilhelm Conrad Röntgen, discoverer of X-rays.

established that his X-rays were capable of penetrating many opaque substances, a fact that was quickly utilized as a diagnostic tool in medicine, used both for radiation treatment and as a way of examining the interior of the body (Fig. 7.5).

In 1896, Henri Becquerel (1852–1908), a French physicist, found that compounds of the element uranium, which had been known about since 1789, spontaneously emit energy which had penetrating powers similar to Röntgen's X-rays. Subsequently he found this energy could be deflected by a magnetic field.

Working in France at the same time as Becquerel were the Curies, Pierre and Marie (Fig. 7.6). Originally from Poland, Marie Curie was interested in Becquerel's findings and began investigating natural substances that showed similar effects. Pitchblende, which is an impure form of the mineral uraninite, the most common ore of uranium, was found to emit more energy than pure uranium. Her suspicion was that there was an unknown substance contained within the pitchblende and working with Pierre she searched for it. In 1898 they discovered two new elements, radium and polonium, which

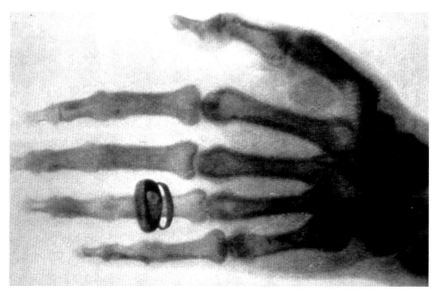

Figure 7.5 An early X-ray picture by Wilhelm Röntgen of his wife's left hand (hence the wedding ring), 1896..

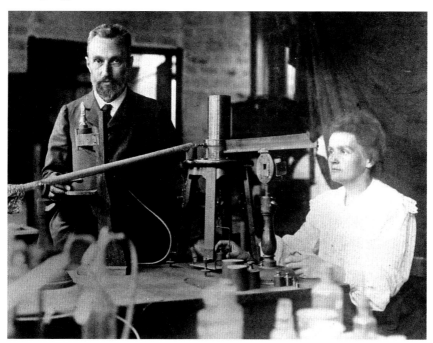

Figure 7.6 Pierre and Marie Curie, originators of the term radioactivity. © musée CURIE

spontaneously emitted energy. Marie Curie coined the term 'radioactivity' to describe such properties and the Curies and Henri Becquerel shared the 1903 Nobel Prize in Physics for their discoveries.

The discovery of X-rays and radiation was to have profound effects on many aspects of geology within a short time. The developments in physics and chemistry not only enabled geologists to respond to Kelvin's arguments based on his cooling Earth model, but radioactivity was eventually to provide a robust method for calculating the absolute ages of rocks and minerals and so finally answer the question of how old was the Earth. Before that happened, however, John Joly, who unsuccessfully attempted to gauge the age of the Earth by measuring salt accumulation in the oceans, thought he had found another way by using the newly discovered phenomenon of radioactivity.

John Joly and the pleochroic halos

The discovery of X-rays and radioactivity may have made the scientific headlines of the time, and rewarded their discoverers with Nobel Prizes, but at more mundane levels science was making progress in other fields. One of these was in microscope technology and analytical methods. By the 1860s a technique had been developed that revolutionized the understanding of the minerals and textures of igneous, sedimentary and metamorphic rocks. It was discovered that very thin slices of most rocks were transparent when light was passed through them. With the improvement in microscope quality that came with advances in lens preparation, this meant that the nature of grain boundaries within even fine-grained rocks could be more readily investigated. To use this property a thin sliver of rock is cut from the sample with a diamond saw. It is then mounted on a glass slide and ground down using progressively finer abrasive grit until the sample is 30 μm thick.

Even more information could be obtained using polarizing lenses in a microscope. Most people are familiar with polarizing sunglasses. They work by allowing only light vibrating in one direction to pass through the lenses, thus cutting out the randomly orientated reflected rays which are uncomfortable for the eyes. If the rock slice (called a thin section) is placed between two such polarizing lenses set at right angles to each other, then the optical properties of the individual minerals making up the rock alter the colours and intensity of the light as seen by the person looking down the microscope. As each mineral type has unique optical properties it means that under the microscope it is comparatively easy to identify the individual minerals that the rock is composed of. It is also possible to see details of the texture of the rock and the relationships between the various minerals.

Figure 7.7 is a view of a thin slice or section of granite, viewed down the microscope and through crossed-polarizers. Three distinct minerals are visible, separated by their different optical properties. The mineral showing up as brown or green is biotite, a form of mica, while the darkish grey mineral showing parallel

Figure 7.7
Photomicrograph of granite in cross-polarized light. The field of view is roughly 4 mm. Courtesy of J.A.Gamble.

stripes in places is the mineral feldspar, a complex silicate mineral containing potassium, sodium and aluminium as well as silicon. The pale grey/white mineral lacking in any stripes is the mineral quartz, a characteristic mineral of granitic rocks. The interlocking jigsaw fit of the mineral grains, typical of an igneous rock, is also clearly seen. It was in granites from Ireland and using these microscope techniques that John Joly developed an innovative attempt to estimate the age of the Earth. Against the background of rapidly moving developments in physics and chemistry Joly felt that radioactivity could hold the answer to the question of the Earth's antiquity. By 1905 radioactive minerals were reported as occurring widely in many rocks and by 1907 Joly was investigating the significance of some features he had observed in the Leinster Granite in Ireland. When the crystals of biotite in the granite were examined in detail he discovered they had small dark rings in them (Fig. 7.8A and 7.8B).

Joly realized that each of the dark circles, which came to be known as pleochroic halos, had a crystal of the mineral zircon in the centre of the circle. These crystals were inclusions in the biotite. The biotite had crystallized around the earlier formed zircon, completely enclosing it. Pleochroism is a diagnostic property of the mineral biotite whereby it changes the intensity of colour, in this case various shades of brown, as the crystal is rotated under the microscope. In Figure 7.9B the dark ring in the brown biotite has a colourless mineral at its centre, this is the zircon inclusion. Zircon had earlier been shown to have radioactive properties and the dark rings were in effect radiation damage to the biotite crystal caused by the emission of energy from the zircons. Joly concluded that the width of the halo was related to the type of radiation involved and the intensity of colour of the ring was governed by the amount of time elapsed since the zircon had been included in the biotite. He hoped that it would be possible to equate the radius of the halo with the age

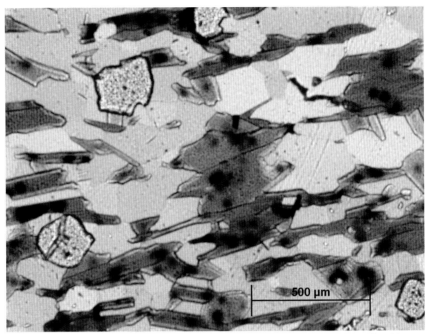

Figure 7.8A Black pleochroic halos in brown biotite crystals. Courtesy of J.A.Gamble.

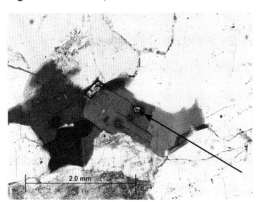

Figure 7.8B Pleochroic halos (black, arrowed) surrounding zircon inclusions in biotite (brown). Courtesy of J.A.Gamble.

of the rock and thus use it to develop a chronology. However, there proved to be too many inconsistencies in the dimensions of the halos when compared with the likely ages of the rocks examined. While he found examples where the relationship between age and halo radius seemed to hold, he then discovered Archaean rocks from Norway, very old rocks, which possessed small halos. By the late 1920s the accuracy of Joly's measurements were being questioned and the method was subsequently discarded.

It was, however, an imaginative and innovative way of using the new phenomenon of radioactivity to assess the age of minerals and rocks and along with his earlier ideas on oceanic sodium accumulations represents a significant contribution to the quest for an absolute geological timescale. The particular phenomenon of radioactivity would indeed be used to construct an absolute timescale, but not in the way envisaged by Joly.

Radiometric dating

The discovery of X-rays by Röntgen in 1895 and the subsequent realization by the Curies in 1903 that the new element radium they had isolated for the first time was a source of spontaneous heat – radium was shown to be capable of melting its own weight of ice in a day – was a life-line to the geological community. They were struggling to come to terms with the conflicting demands of the time required for evolutionary processes and the constricted timetable allowed by Kelvin's calculations of the age of the Earth. Since radium could be proved to produce heat in the laboratory it could undoubtedly produce heat in its natural environment in the Earth's crust and mantle.

The next major player in the elucidation of radioactivity and its application to the measurement of geological time was Ernest Rutherford (Fig. 7.9), a New Zealander, born in 1871 to emigrants from England and bright enough to win a scholarship to Cambridge.

Rutherford developed a general theory of radiation, the disintegration theory, which stated that the atoms of radioactive bodies are unstable and every second a certain fixed proportion of them break up with the release of

Figure 7.9 Ernest Rutherford (1871–1937) in 1892.

energy in the form of alpha-particles. These had been discovered by Henri Becquerel when he found that the energy coming from certain uranium compounds could be deflected by a magnetic field, thus proving that particles were involved. In a landmark book published in 1904, *Radioactivity*, Rutherford explained that radioactivity results from the spontaneous disintegration of an unstable element into another element, with the emission of energy often in the form of sub-atomic alpha- or beta-particles. This new element may decay further until a stable element is finally created. The sequence of transformations is now known as a 'radioactive decay series'. This disintegration theory came to provide the basis for the numerical quantification of geological time.

Rutherford also showed that the rate of radioactive decay is constant and is not affected by any external factors. He also speculated that the end-product of radioactive decay was the gas helium, and that if the amount of helium trapped in the minerals could be measured, then estimates could be made for the ages of the minerals themselves. Following on from Rutherford's speculation, in 1905 Robert John Strutt, later to be Baron Rayleigh, successfully analysed the helium content of a radium-bearing rock and determined its age to be 2 byr – the first successful application of a radiometric technique to the calculation of a radiometric date for a rock or mineral. The technique was very much in its infancy and there were clearly problems such as the loss of helium from the system over time, but it cannot be regarded as anything less than a great scientific advance. The scale of the breakthrough was confirmed in the same year from across the Atlantic by an American chemist Bertram Borden Boltwood. He had speculated that since lead is always found with ores of uranium it might be another of the decay products in the uranium series. On this assumption he calculated the ages of 43 minerals, getting a range from 400 myr to 2.2 byr. While these figures were comparable to those obtained by Strutt, they were clearly of a higher order of magnitude than the figures obtained by the use of sedimentation rates or heat flow.

As so often in the preceding 19th century, the early part of the 20th century was an interesting time to be an Earth scientist. The scale of the figures proposed by Strutt and Boltwood were so large that many in the geological community were having difficulty adjusting to the expansion of geological time from Kelvin's low of 20 myr to literally thousands of millions of years. As well as that, some workers such as John Joly criticized the basis of the method, arguing whether the decay rates of radioactive minerals were as constant over geological time as had been suggested. At this juncture we have the first appearance on the geological stage of a man whose contribution to the Earth sciences over much of the 20th century was to be of major significance and lasting influence.

Arthur Holmes: geochronologist

Arthur Holmes (1890–1965, Fig. 7.10) was an English geologist who was an important innovator in the development and propagation of possibly the two most fundamental revolutions and paradigm shifts in world geology in the 20th century. These were the use of radiometric dating techniques to grasp the extent of geological time and unravel Precambrian time, and the development of the unifying concept of plate tectonics. His textbook, *Principles of Physical Geology*, was a constant presence on the bookshelves of undergraduate geology students for over 50 years after the first edition was published in 1944.

Holmes was born near Newcastle upon Tyne in the north-east of England and spent his childhood at the nearby town of Gateshead. In 1907 he took up a scholarship at what is now Imperial College in London. He started off reading physics but took geology courses in his second year in a move that would influence his later career path. After graduation he had the opportunity to stay at Imperial College, working in Strutt's laboratory on radiometric dating problems, from where he published his first scientific paper at the precociously early age of 21 in 1911. This was in no less a journal than the *Proceedings of the Royal Society*. This pioneering paper was on the use of the uranium–lead decay series in calculating the age of a range of rocks, including an age of 1,640 myr for Precambrian rocks from Ceylon (now Sri Lanka) and giving an age of 370 myr to a nepheline syenite rock from Norway. The data in the paper provided numerical ages for rocks from several periods of the Palaeozoic, as well as from the Precambrian. For example Carboniferous-aged material was calculated as 340 myr, the Norwegian nepheline syenite was considered to be Devonian (370 myr), Ordovician/Silurian rocks were dated at 430 myr, and the Precambrian specimens analysed ranged in age from 1.025 to 1.64 byr. Holmes further consolidated his reputation in the field of geochronology by publishing a book entitled *The Age of the Earth* in 1913.

Figure 7.10 Arthur Holmes in 1912.

In this book he reviewed critically the estimates of the age of the Earth based on various astronomical, geological, geothermal and radiometric processes, opting for radiometric methods as the way forward. The significance of the figures produced by Holmes was that for the first time absolute age dates were proposed for rocks within a stratigraphical or relative age context. This remarkable book and the preceding Royal Society paper meant that the relative geological timescale, the stratigraphical column, which had evolved over a period of some 200 years since the ideas of people like Nicolas Steno had been put forward, could now be quantified. As well as having superpositional significance the stratigraphical column now also had chronological significance. These publications by Holmes at an early stage in his career can be considered as the first real steps towards a graduated geological timescale with absolute dates, a process that continues today as analytical procedures and equipment become ever more sophisticated.

Radiometric dating: principles and methodology

In the earlier discussion of stardust reference was made to isotopes – versions of elements which are heavier or lighter than normal due to differing numbers of particles called neutrons in their atomic structure. The discovery of isotopes by Frederick Soddy in 1913 was to have far-reaching consequences not only for the measurement of absolute time and the age of the Earth but also in helping us understand geological and environmental processes such as the chemical evolution of igneous rocks and the reconstruction of past climates. However, around the time of their discovery, isotopes were to play a key role in the absolute measurement of geological time and the growth of the new branch of science which came to be known as geochronology.

Frederick Soddy was a research associate of Ernest Rutherford and a co-author of their disintegration theory of radioactive decay in the early 1900s. In 1913 he demonstrated that certain elements could exist in more than one form which are identical to each other chemically but have slight differences in their atomic weights. The ages calculated by Holmes for his 1911 paper were based on the decay series uranium–lead, and it is now known that both uranium (U) and lead (Pb) exist in different isotopic forms which means that Holmes' calculations could not have been completely accurate since he was not aware of these isotopic variations.

The mass number of an element is the number of protons and neutrons in the atomic structure of the element. Atoms of the same element contain the same number of protons in their atomic nuclei, but isotopes of that element will have a different number of neutrons and therefore a different mass number – hence ^{238}U and ^{235}U for example.

Uranium-238 decays to lead-206 (^{206}Pb), and the rate at which this decay takes place is such that half of the atoms of uranium will have decayed after 4,470 myr – this is called the half-life of the isotope. The form of uranium with a mass number of 235 (^{235}U) decays to form lead with a mass number of 207 (^{207}Pb) with a half-life of 704 myr. Other examples of decay series are: ^{40}K (potassium) decays to ^{40}Ar (argon) with a half-life of 1,193 myr and ^{87}Rb (rubidium) decays to ^{87}Sr (strontium) with a half-life of 48,800 myr.

In the pioneering days of radiometric dating there were uncertainties over isotopic variations of the elements, the various decay series present were not fully understood and there was disagreement over the extent to which decay rates were constant. Clarification of many of these issues had to await the development of better analytical techniques and equipment. The mass spectrometer became important from the mid-1920s onwards, allowing the identification of isotopic forms and the determination of relative abundance.

In considering the theory behind radiometric dating it is worth reviewing the words of Holmes himself in that ground-breaking paper in 1911, explaining why he was analysing minerals that he knew contained uranium and lead:

> Such minerals may be regarded as storehouses of the various series of genetically connected radioactive elements. In them the parent element slowly disintegrates, while the ultimate products of the transformation gradually accumulate. The analysis of these minerals ought, then, in the first place, to disclose the nature of the ultimate product of each series; secondly a knowledge of the rate of formation of this product, and of the total quantity accumulated, gives the requisite data for calculation of the age of the mineral.

This explanation of the decay process and the basis of calculating the age of the mineral are essentially unchanged from the way in which a radiometric age is calculated today. Atoms of elements exhibiting radioactivity are inherently unstable and over time decay to form a new element – the process is depicted as 'parent atoms' decaying to form 'daughter atoms'.

The radioactive elements such as uranium used in the calculation have been incorporated into minerals such as zircon which have crystallized in, for example, cooling igneous rocks like basalt or granite. This incorporation marks the start of the radiation clock and the element, for example uranium, begins to decay at a steady and predictable rate. By measuring the number of parent atoms left in the rock and comparing it with the number of daughter atoms, it is then possible to calculate the duration of the process. By implication this gives the age of the mineral and therefore the age of the rock containing it.

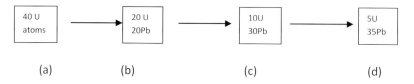

40 U atoms	20 U 20Pb	10U 30Pb	5U 35Pb
(a)	(b)	(c)	(d)

The process is depicted diagrammatically above with the decay of uranium atoms to lead. For convenience the half-life is taken as a billion years (1 byr) and there are initially 40 uranium atoms when the decay process starts (a). After the 1st half-life (1 byr) half of the uranium atoms have decayed to lead, giving equal numbers of lead and uranium atoms in the sample (b). If the decay continues for another 1 byr (two half-lives) there are 10 uranium atoms and 30 lead atoms and the rock is 2 byr old (c). If the decay process continues for another half-life period (three half-lives) there would be 5 uranium atoms, 35 lead atoms and the rock would be 3 byr old. The total number of atoms present in the system does not change with time, merely the ratio of parent atoms to daughter atoms changes.

Figure 7.11 shows this change. After one half-life the number of parent and daughter atoms are equal. After this the number of parent atoms continues to fall while the number of daughter atoms correspondingly rises.

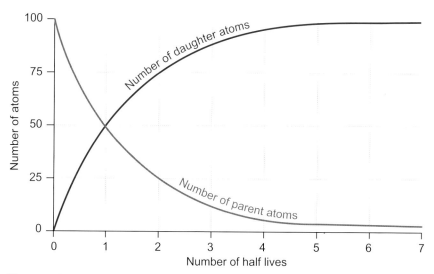

Figure 7.11 Graph showing the rate of increase of daughter atoms and the rate of decline of parent atoms with time.

The calibration of the stratigraphical column

The development of radiometric dating techniques and the absolute age dates they generated became a powerful tool in calibrating the stratigraphical column which was based on relative ages. For such a calibration there has to be integration between the stratigraphical column, which is based largely on sedimentary rocks and the fossils contained within them, and the absolute age dates which are obtained mainly from igneous and metamorphic rocks.

Radiometric dating measures an event in the history of the rock being analysed. In the case of a mineral from an igneous rock the event is the cooling of the rock below a certain temperature resulting in the radioactive element being trapped within the rock. For metamorphic rocks the date will be when the rock cooled after it had been recrystallized in the solid state after metamorphism. It is possible in some cases to date sedimentary rocks (which are mainly composed of eroded fragments of older rocks) if new minerals have formed during diagenesis which is the process whereby, for example, loose sand grains are compressed and cemented to form sandstone. To use absolute age dates from igneous and metamorphic rocks we need to use the fourth of Steno's original laws of stratigraphy (his first, second and third laws concerning superposition, original horizontality and lateral continuity were introduced in Chapter 6). This fourth law is the law of cross-cutting relationships, as illustrated below (*see* fig 6.1C and Fig. 7.12).

In this simple case the layers of sediment must be older than the dyke which intrudes them at 100 myr, but younger than the lava which was erupted 125 mya. On the basis of this evidence it can be stated that the sediments were deposited at some time within this time interval. This technique is known as bracketing and using it has enabled the calibration of the whole stratigraphical column.

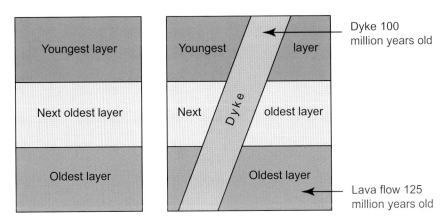

Figure 7.12 Law of cross-cutting relationships. The whole sequence of rocks is older than the dyke.

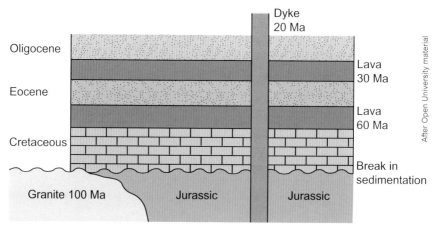

Figure 7.13 Calibrating the stratigraphical column using radiometric dates.

To show how relative dating and numeric/absolute dating methods are integrated, it is useful to examine a theoretical example first. The cross-section (Fig. 7.13) shows rocks which are known from the fossils they contain to range in age from the Jurassic system, through the Cretaceous and into Eocene and Oligocene subdivisions of the Palaeogene. There are a number of igneous rocks associated with this sedimentary sequence and they have had their ages calculated radiometrically. Using these dates and the cross-cutting relationships the rocks display it is possible to integrate the relative and absolute ages for the succession.

Starting with the base of the sequence, the Jurassic is intruded by the granite which is 100 myr old, which means that the Jurassic must be older than 100 myr. From the information available it is not possible to determine its maximum age. The Cretaceous which overlies the Jurassic has been deposited on top of the granite intrusion, so therefore must be younger than 100 myr, but it is overlain by a lava flow which is 60 myr old and this gives the minimum age for the Cretaceous. Immediately following this lava flow are the sediments of the Eocene and Oligocene. The Eocene is bracketed by a second lava flow which is 30 myr which means the Eocene is less than 60 but more than 30 myr old. The Oligocene is less than 30 myr old because it is underlain by a lava flow of this age, but along with the rest of the sequence it is cut by a 20 myr old dyke, so it must be older than 20 myr.

In summary:

- the whole sequence is older than 20 myr;
- the Oligocene is older than 20 but younger than 30 myr;
- the Eocene is older than 30 but younger than 60 myr;
- the Cretaceous is older than 60 but younger than 100 myr;

- the Jurassic is older than 100 myr, but its maximum age cannot be determined from the available evidence.

Using these techniques to integrate the relative and absolute ages of thousands of rocks in thousands of successions around the world, a geological timescale has been produced that provides a temporal framework against which all geological events can be referenced. If one person could be singled out as having contributed most to this state of affairs, then that person would be Arthur Holmes who had the vision of using isotopes as a measure of geological time and followed it to a successful conclusion.

Unravelling the Precambrian

The geological column was constructed using layered rock sequences containing a fossil record. The term Precambrian was originally applied to those rocks which were older than rocks of Cambrian age, which by the end of the 19th century were considered to be the oldest fossil-bearing sediments on Earth. Problems arose, however, when it began to be realized that a considerable period of Earth history occurred before the arrival of shell-bearing organisms. This was clearly a major limitation on the complete understanding of Earth processes based on whether or not organic remains are able to be preserved by means of their mineral shell. It meant that the geological timescale as it then existed applied only to a relatively short period of the total extent of geological time.

The stratigraphical column constructed on this basis usually had a small section at the base labelled as Precambrian, while the bulk of the column was made up of the younger Palaeozoic, Mesozoic and Cenozoic eras. This representation gives a totally false impression of the significance of Precambrian time to Earth history. The true extent of Precambrian time compared with that of the younger Phanerozoic can be seen in Figure 7.14.

Rather than representing an apparently short period of time before the main action of the Phanerozoic took place (Palaeozoic + Mesozoic + Cenozoic), the Precambrian has been shown to account for almost 90% of the total history of the Earth. The fossil record of the Precambrian is much sparser than the Phanerozoic, not only because hard-shelled life formed had still to evolve but also because many rocks of this age have been heavily metamorphosed which obscures their original form and would destroy any fossils contained within them. Such fossils as do exist in the Precambrian such as stromatolites are of limited use for stratigraphical purposes, so the calibration/subdivision of the Precambrian has depended on the techniques of radiometric dating combined with the interpretation of cross-cutting relationships.

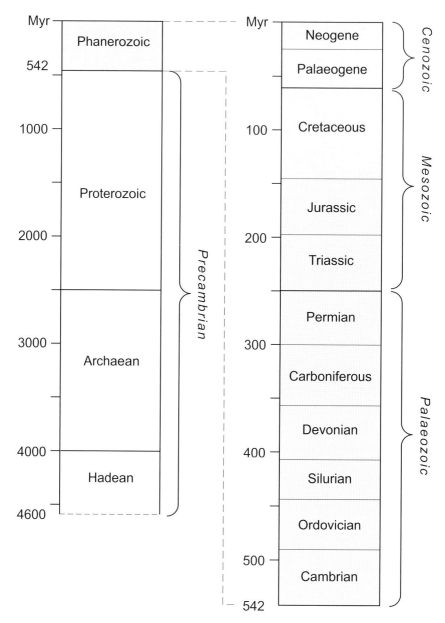

Figure 7.14 The stratigraphical column: the left-hand section shows the true-scale extent of Precambrian time. After Upton.

The subdivision of the Precambrian

The 19th-century constructors of the stratigraphical column based the Cambrian on the earliest known fossils available at the time, and assumed that everything below this level was essentially devoid of life. It was therefore labelled 'Pre-Cambrian'. The definition was based on the first appearance of macroscopic hard-shelled animals in abundance. Radiometric dating has now given a measure of how long this period extends for, and since the beginning of the 20th century research has shown that the Precambrian, far from being largely devoid of life forms, in fact had a diverse fauna. The study of this has shed much light on the evolutionary development of life from very early times in the history of the Earth. The subdivision of the Precambrian is therefore as important as the calibration of the younger Phanerozoic section of the stratigraphical column. The Cambrian, Ordovician and Silurian periods were largely defined on the basis of marine invertebrates such as brachiopods with shells of similar material to modern cockles and mussels, or trilobites with hard exoskeletons like modern crabs or lobsters.

Fossils found in the Ediacara Hills of South Australia, discovered first in the 1940s, are older than those used to define the Cambrian, and they represent the oldest multicellular organisms. They occur in a range of shapes including circular discs with internal radial arrangements or concentric ribs as in *Cyclomedusa plana* (Fig. 7.15) and are thought to be possibly similar to modern-day jellyfish, sea anemones or annelid worms. The Ediacaran rocks thus give us a good look at the first animals to live on Earth in the Proterozoic Eon.

The Ediacaran Period, based on this distinctive fauna, was ratified officially in 2004 by the International Union of Geological Sciences (IUGS), making

Figure 7.15 *Cyclomedusa plana*, an Ediacaran fossil from the Flinders Ranges, Australia.

it the first new geological period to be declared in 120 years. It immediately precedes the Cambrian Period, the first period of the Phanerozoic (visible life) Eon and is the last geological period of the Proterozoic (early life) Eon. The age of the Ediacarian Period is estimated to be 635–541 myr, based on data from a number of localities throughout the world where the fauna has been recognized.

Figure 7.16A shows a layered stromatolite, produced by the activity of ancient cyanobacteria. These are the oldest known fossils, with some examples dating to around 3.5 bya. Cyanobacteria are water-living and capable of photosynthesis and can therefore produce their own food. They are sometimes referred to as 'blue-green algae', although strictly speaking they are not related to algae. The layers of the stromatolite were produced as calcium carbonate precipitated over the growing mat of bacterial filaments; photosynthesis in the bacteria depleted carbon dioxide in the surrounding water, initiating the precipitation of the calcium carbonate. The minerals, along with grains of sediment precipitating from the water, were then trapped within the sticky layer of mucus that surrounds the colonies of bacteria. The bacteria continued to grow upwards through the sediment to form new layers, over and over again.

Figure 7.16A Fossil stromatolite, India. © Dr Ajay Kumar Singh/ Shutterstock

Stromatolites reached a peak about 1 bya and were widespread in the Proterozoic, ecologically important as the first reefs. By the end of the Proterozoic their abundance decreased rapidly, possibly due to grazing by more advanced life forms.

The cyanobacteria associated with stromatolites played an important role in shaping the course of evolution and ecological change in the early stages of Earth's history. We now depend on oxygen for survival, but the early atmosphere on Earth had a very different chemical make-up from now and was unsuitable for advanced life. Photosynthesis by cyanobacteria during the Archaean and Proterozoic Eons began the process of oxygenation which led eventually to the formation of the protective ozone layer and enough oxygen in the atmosphere to allow the development of life forms more complex than bacteria.

Stromatolites have proved to be among the great survivors in Earth history and in localities such as Shark Bay in Western Australia modern forms are growing in exactly the same way as they have done for 3.5 byr (Fig. 7.16B).

The Precambrian therefore, far from being a fossil-free zone as originally thought by those geologists compiling the first stratigraphical columns, has an extensive fossil record. This extends from the earliest cyanobacteria at about 3.5 bya to the diverse life forms of the Ediacaran Period from 635 to 541 mya, marking the end of the Proterozoic Eon. Following this was the appearance of animals with hard shells, the Cambrian explosion, and the first period of the Phanerozoic.

Figure 7.16B Stromatolites in Shark Bay, Western Australia. © Rob/Bayer/ Shutterstock

The age of the Earth (or the start of the Precambrian)

The age of the Earth is 4.54 ± 0.05 byr (4.54×10^9 years $\pm 1\%$). This age is based on age dates obtained from meteorites which are considered to represent the primitive material from which the solar nebula was formed. Geological samples from Earth are not considered reliable sources for a direct date of the formation of the Earth because of its early separation into the separate layers of core, mantle and crust, and then its subsequent plate tectonic history which involved the formation and destruction of ocean basins and the movement of continental masses around the surface of the Earth. The oldest rocks on Earth are found in the metamorphic foundations of the continental areas and are all from the Hadean Eon (the first eon, see below) of the Precambrian. These are regions such as the Canadian Shield, the Pilbara Craton of Western Australia, the Baltic Shield and the Congo and Zimbabwe Cratons in Africa. A craton is the old, geologically stable core of a continental area, often in the interior parts of tectonic plates. They are characteristically composed of very old crystalline metamorphic rocks known as basement rock.

The age of the Earth obtained this way is consistent with the ages currently obtained for the oldest known rocks on the Earth's surface. The oldest such minerals analysed to date are small crystals of the mineral zircon, the same mineral responsible for the pleochroic halos in John Joly's observations discussed earlier in the chapter. These are from the Jack Hills of Western Australia and are at least 4.04 byr old. Their existence implies that at this early stage in Earth history there was a stable crust.

The zircons dated from the Jack Hills are detrital, that is they occur as fragments of a pre-existing rock. They are found in a metamorphosed sedimentary rock, in this case a metamorphosed conglomerate, dated at around 3.0 byr, but the zircons themselves had crystallized at an earlier stage (about 4.04 byr) before being eroded from their original rock and then deposited in the conglomerate.

Zircon is a very hard, dense and durable mineral which is also chemically very stable. These properties mean that it is a common residual mineral, as when the parent rock is being broken down by the forces of physical and chemical weathering minerals such as zircon will survive as individual and largely unaltered grains and will be incorporated in the next cycle of rock formation. Diamond is another good example of a residual mineral; like zircon it is also hard, relatively dense and chemically very stable. Many of the world's diamonds are found as detrital grains in former river gravels.

Zircons crystallize in an igneous rock such as granite and at that stage incorporate some of the radioactive element uranium. Because of their durability they are capable of passing through a series of geological episodes such as the rifting and merging of continents. These successive stages can be recorded as new layers of the mineral which are added around the original crystal, like growth rings in a tree. Each of these layers represents a separate crystallization event which may have lasted tens or hundreds of millions of years. Although the zircon grain may be a few millimetres in size, the analytical techniques now available allow the individual growth zones of the zircon to be analysed using a thin beam of electrons capable of measuring the composition of a spot a few micrometres across. It is also possible to calculate an age for the individual growth zones so that we can build up a detailed history from a single grain.

The gathering of data at this level of detail has greatly enhanced our understanding of the processes that operated in the very early days of the Earth and how the differentiation into the core, mantle and crust occurred and how the earliest continents evolved.

Summary of the subdivision of the Precambrian

The term Precambrian is now recognized as a general one which is made up of three eons. From oldest to youngest they are the Hadean, the Archaean and the Proterozoic. The Earth probably coalesced from material in orbit around the Sun about 4,500 mya and probably had a stable crust by around 4,400 mya. Age dates recorded from meteorites give a similar age to this.

Using available radiometric dates the Hadean is from 4,600 to 4,000 mya, the Archaean from 4,000 to 2,500 mya and the Proterozoic from 2,500 to 541 mya. The Proterozoic is divided into three eras, the Palaeoproterozoic, the Mesoproterozoic and the Neoproterozoic (see the table below), with the Ediacaran Period the youngest period within in the Neoproterozoic.

Precambrian 4,600 to 541 mya	Proterozoic 2,500 to 541 mya	Neoproterozoic 1,000 to 541 mya
		Mesoproterozoic 1,600 to 1,000 mya
		Palaeoproterozoic 2,500 to 1,600 mya
	Archaean 4,000 to 2,500 mya	Neoarchaean 2,800 to 2,500 mya
		Mesoarchaean 3,200 to 2,800 mya
		Palaeoarchaean 3,600 to 3,200 mya
		Eoarchaean 4,000 to 3,600 mya
	Hadean 4,600 to 4,000 mya	

There has been an alternative proposal to divide the Precambrian into eons and eras on the basis of stages of planetary evolution rather than the current scheme based on numerical ages. This would have the advantage that it would rely on events that are recorded in the stratigraphic record and such events could be demarcated by internationally agreed stratigraphic sections. These would serve as markers for particular boundaries on the geological timescale.

Four such natural eons can be recognized:

1. **Accretion and differentiation:** The period of planetary formation.
2. **Hadean eon:** A period of heavy bombardment by meteoritic material from about 4,510 mya.
3. **Archaean eon:** A period defined by the first crustal formations until the deposition of banded iron formations, as atmospheric oxygen levels increased. Banded iron formations (BIFs) are distinctive sediments, usually Precambrian, consisting of thin layers of blackish iron ore alternating with iron-poor shales, often red in colour. These were relatively common in the early history of the Earth but are now rare and are thought to be linked to increased oxygen levels generated in the atmosphere by cyanobacteria activity. BIFs occur from about 2,400 mya to around 1,800 mya. This radical change in the composition of the atmosphere is a good example of an irreversible change in Earth history – an example of time's arrow.
4. **Proterozoic eon:** The period of modern plate tectonics and the first animals (the Ediacaran fauna).

Precambrian supercontinents

Since plate tectonic processes began the pattern of plate movement has caused the formation and break-up of continents over time and occasionally this led to the formation of a supercontinent where most or all of the continents had accreted. Reference has already been made in Chapter 5 to the supercontinent Rodinia, thought to have formed about 1 bya and broken up about 600 mya. It was replaced by the next supercontinent Pangaea whose eventual break-up about 200 mya has given us the distribution of continents we are familiar with today. We will return to this supercontinent cycle and its implications for Earth history in Chapter 9.

Chapter 8

Archaeological time

Archaeology in essence then is the discipline with the theory and practice for the recovery of unobservable hominid behavior patterns from indirect traces in bad samples...

David Clarke (1937–76) Archaeology: the loss of innocence, *Antiquity* 47, 1973

Archaeological knowledge has been expanding rapidly over the last few decades and much of this growth has been based on a range of dating techniques that has allowed the construction of a reliable absolute timescale in much the same way as happened in geology with the advent of radiometric dating. In addition to radiometric dating using isotopes of the element carbon, archaeological dating now uses techniques such as tephrochronology, which is the study of volcanic ash layers, dendrochronology, the study of tree rings, and magnetochronology, the study of variations in the Earth's magnetic field.

Archaeology is the study of past time through the examination of the material remains left as the result of activity by human beings. Often this involves excavation – the systematic uncovering of a site layer by layer, and recording in detail the soils, artefacts, human remains, evidence of buildings and any environmental information found in successive layers. Archaeologists share with geologists the desire to work out the sequence of past events and like geologists they do so using the stratification of layers of material. As in a succession of rock layers, the oldest material will have been deposited at the bottom of the sequence, while younger material will be found in progressively higher layers. In the same way as radiometric dating allowed the calibration of the stratigraphical column, so the new dating techniques mentioned above have enabled the construction of an absolute timescale in archaeology.

As indicated by the rather ironic quotation from David Clarke at the head of the chapter, another similarity between geology and archaeology is the knowledge that the evidence is always incomplete. Just as fossil evidence can be lost or entire sequences of rocks can be eroded, so archaeological sites can be destroyed by succeeding generations, or stone or metal implements can be reused, so that perhaps only a small fraction of the original evidence is preserved. All of this means that precise and accurate dating methods are vital in any assessment of archaeological evidence.

Tephrochronology

The term *tephra* (from the Greek meaning ashes) is used to describe all the solid, fragmented material produced from a volcano during an eruption. The fine portion of this material, described as ash, can travel long distances before falling to earth and forming a deposit. While the understanding of the formation and subsequent distribution of volcanic ash clouds is important for their impact on climate and the environment, they are of interest to us in this book as a technique in geochronology. The idea of using layers of tephra as a chronological tool, tephrochronology, was first developed in Iceland in the 1940s. It is based on the premise that the tephra layers represented an isochronous or equal-time marker horizon within a succession, which could then be mapped across a wider area. This is analogous to the use of fossils for correlation, as discussed earlier (see Chapter 6). It is also possible to recognize tephra deposits in the oceans, so enabling the correlation of marine and terrestrial deposits.

Tephrochronology therefore can be used to build a chronological framework within which environmental and archaeological events and processes can be placed. It is the archaeological record which is of particular interest in this chapter. To fully appreciate the value of tephra deposits in measuring time in the historical past it is necessary to understand the processes whereby volcanoes produce and distribute ash particles and how the unique properties of each ash fall – such as its chemistry, grain size, texture and density – can be measured and linked back to its parent volcano and eruption.

The awesome power exhibited in the eruption of the volcano Mount Pinatubo in the Philippines (Fig. 8.1) is due to a type of eruption described by volcanologists as Plinian, after the first person to describe the phenomenon, Pliny the Younger. He and his uncle, Pliny the Elder, witnessed the eruption of Vesuvius in 79 CE that destroyed the Roman towns Pompeii and Herculaneum. While attempting to evacuate a family from the shore during the eruption the older man died, and his nephew described the eruption and surrounding events in a letter written to a friend.

Plinian eruptions are the largest and most violent of volcanic eruptions and are characterized by columns of gas and ash extending far up into the atmosphere, up to 50 km high. These ash clouds can travel hundreds or thousands of kilometres away from their source. The recent (2010) eruption of the Icelandic volcano Eyjafjallajökull, which caused so much disruption of air travel in Europe and North America, was a Plinian eruption. Although on a very minor scale

Figure 8.1 The 1991 eruption of Mount Pinatubo in the Philippines, the largest eruption on Earth since 1912. USGS image.

Figure 8.2 The eruption cloud of Eyjafjallajökull, Iceland, 2010. © J.Helgason/ Shutterstock.

compared with some other examples we are about to consider it serves very well to illustrate the characteristics of that type of eruption (Fig. 8.2) and the distance the ash cloud can travel (Fig. 8.3).

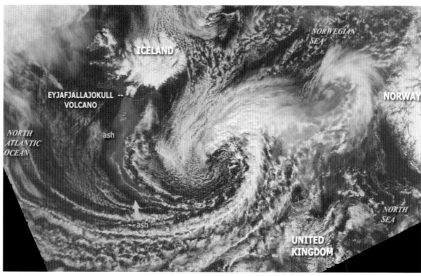

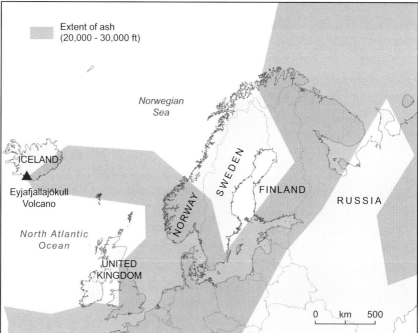

Figure 8.3 Distribution of the ash cloud over Europe a few days after the eruption of Eyjafjallajökull in 2010. Image NASA. Map after BBC News.

The 1991 eruption of Mount Pinatubo was the biggest recorded eruption on Earth since the eruption of Mount Tambora on Indonesia in 1815. The Tambora eruption was probably some ten times bigger than the much better known eruption of Krakatoa in 1883. The aftermath of the Tambora eruption was global cooling caused by the amount of ash and gas in the atmosphere, which in turn led to widespread harvest failure and famine across the world. The year 1816 is often referred to as 'The Year Without a Summer'. The greater notoriety of Krakatoa was probably due at least in part to the better communications that existed at the end of the 19th century compared with the early 1800s, when news of the effects and scale of the eruption would probably have taken 6 months to reach Europe.

The comparative size of some historically recorded eruptions is given below, with the magnitude of the eruption based on the estimated volume of material ejected:

- Pinatubo 1991: 5 km^3;
- Krakatoa (Krakatau) 1883: 9.2 km^3;
- Tambora, Indonesia, 1815: 100 km^3;
- Laki, Iceland, 1783: 14 km^3;
- Vesuvius 79 CE: 4.2 km^3;
- Thera (Santorini): 3.6 kya: ~60 km^3;
- Toba, Indonesia, about 74 kya: ~2,800 km^3.

As can be seen from this list, the Toba eruption approximately 74,000 years ago is one of the largest known to have occurred on Earth, ejecting nearly 3,000 km^3 of magma and causing worldwide climatic and environmental impacts.

Volcanoes and archaeology

It is a paradox that volcanoes which are generally thought of as forces of destruction have in fact a number of very positive aspects. The reason that the slopes of active volcanoes are frequently densely populated is because the ash products from eruptions often provide very fertile soils after a relatively short weathering period. Many civilizations originated on the slopes of volcanoes and around 500 million people around the world continue to inhabit potentially hazardous volcanic environments. The lure of fertile soils, geothermal energy, valuable ore and mineral deposits and the tourism opportunities that derive from some of the most spectacular landscapes found anywhere outweigh the potential hazards.

The other role played by volcanoes, which is perhaps counter-intuitive, is as a force for conservation and preservation. The speed of engulfment of settlements by the ash clouds means that large areas are quickly buried thus preserving buildings and artefacts from subsequent damage by erosion. The ash layer effectively freezes time at the instant of burial, providing a 'snapshot' of life (and death) in that settlement at the time of the eruption.

The vulnerability of Pompeii to an explosive eruption is clearly demonstrated in Figure 8.4, with the volcano less than 10 km away. The human cost of this type of eruption is poignantly demonstrated by the casts of the bodies of the trapped inhabitants of the city as they were overcome by the choking clouds of searing

Figure 8.4 The ruins of Pompeii with Vesuvius in the background. © Darryl Brooks/ Shutterstock

Figure 8.5 Casts of the bodies of victims buried in the Fugitives Garden at Pompeii.

hot gas and ash (Fig. 8.5). Although the city had been subjected to an increasingly violent bombardment of ash falls, many of the inhabitants had remained in their homes, presumably in the hope that the eruption would die down and they could resume their normal lives.

Nearly 1,700 years before the eruption which buried Pompeii and Herculaneum, and some 900 km to the south-east of Vesuvius, on the Greek island of Santorini (also called Thera, Fig. 8.6), the Bronze Age settlement of

Figure 8.6 Island of Santorini (Thera). © Gyuszko-Photo/ Shutterstock

Akrotiri was buried by an eruption in 1613 BCE which dwarfed the 79 CE Vesuvius eruption. Compared with the 4.2 km³ of ash erupted by Vesuvius, the explosion at Santorini produced 60 km³ of ash in what was one of the largest Plinian eruptions on Earth in the past 10,000 years.

In the second millennium BCE Akrotiri was a thriving sea port on Santorini, trading with the island of Crete some 100 km to the south. Akrotiri was contemporaneous with the Minoan civilization on Crete which dominated the eastern Aegean region between about 2,000 and 1,450 BCE and is characterized by the ruins at Knossos.

The scarcity of valuable items such as gold and jewellery and the complete absence of human remains suggest that (in contrast to Pompeii) Akrotiri's population had sufficient time to evacuate the port before the final eruption which buried it. This final eruption sequence began with the deposition of a thin layer of ash followed by a cataclysmic explosion that generated a huge volume of pumice material. Most of the island today consists of the large caldera which is all that remains of the original volcano.

The pumice (Fig. 8.7) and ash fallout was dispersed by the prevailing winds over a large area of the Mediterranean region, as far as central Turkey. Studies of the distribution of the volcanic ash layer from the eruption indicate that most of the ash was transported eastwards where several tens of centimetres of ash accumulated at distances up to 200 km from the volcano.

Since the eruption of Thera was so much bigger than the eruption of Vesuvius in 79 CE, then so the overall effect on the eastern Mediterranean area was also

Figure 8.7 Minoan pumice deposits, Santorini. © Helen Field/ Shutterstock

much greater. An eruption of this magnitude would have had profound effects such as tsunamis, earthquakes, blankets of pumice on land and on the sea and the ash in the atmosphere would have triggered climatic changes that would have affected a very wide area for a long time after. In addition the eruption left a synchronous layer of tephra over the eastern Mediterranean area which formed a conspicuous and widespread marker horizon for tephrochronology. The eruption products potentially provide important evidence for the dating of events for the period around 3.6 kya when major cultural changes were taking place in the eastern Mediterranean region, many possibly in the aftermath of what was a catastrophic volcanic event.

It is this ability of ash clouds to travel long distances that is the basis of tephrochronology. The technique is based on discrete layers of ash from a single eruption creating a chronological framework within which archaeological events, such as described above, or environmental changes can be placed. Such a 'tephra horizon' has a unique chemical signal or 'fingerprint' that allows the ash deposit to be identified across the fallout area and also can allow the source of the ash to be confirmed. If the volcanic episode has been independently dated, perhaps by historical records, then the tephra horizon becomes a time marker.

In an active volcanic region such as Iceland where there are frequent eruptions it is possible to identify separate tephra horizons linked to each volcanic episode. Figure 8.8 is a section through the soil in southern Iceland showing distinctive tephra layers in the topsoil.

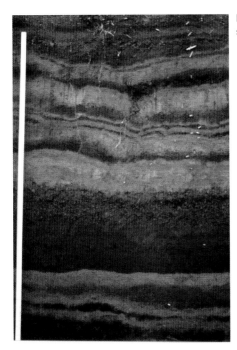

Figure 8.8 Historical tephra layers in a soil profile in southern Iceland.

Icelandic ash in Britain and Ireland

Close scrutiny of the sedimentary record of Britain and Ireland reveals tephra deposits from Icelandic volcanoes. We have already mentioned the most recent, the 2010 eruption of Eyjafjallajökull, but there have been at least ten eruptions of Icelandic volcanoes in the last 1,000 years which have affected Britain and Ireland. One of the most devastating volcanic episodes in historical times was the 1783 fissure eruption of Laki in southern Iceland.

The line of volcanic cones in Figure 8.9 are the remnants of the eruptive fissures that produced the largest outpouring of lava by volume in recorded history. In an 8-month period around 14 km³ of lava was erupted, releasing clouds of ash and toxic gases that killed a quarter of the island's population and half of the island's livestock. In Britain the summer of 1783 was known as the 'Sand Summer', with ash falls lasting almost a year. In addition a sulphurous haze from the millions of tonnes of sulphur dioxide gas released by the eruption was instrumental in the deaths of thousands of people in Britain that year. The Sand Summer was then followed by a severe winter.

Frequent eruptions of the volcano Hekla in southern Iceland have resulted in tephra fallout in Britain and Ireland, the most recent occurring in 2000. One of its prehistoric eruptions, designated as Hekla 4, and dated to 2319±20 BCE (around 4,300 years ago) was one of the largest explosive eruptions in Iceland since the Ice Age ended about 10,000 years ago. Ash from the Hekla 4 eruption

Figure 8.9 The Laki fissure, site of the 1783 eruption, southern Iceland. © Tokelau/ Shutterstock

can be found across Scotland and Scandinavia, where it is an important tephrochronological marker. Evidence from pollen layers (palynology) appears to show a dramatic decline in pine species around this time, probably as a result of climate change brought on by the ash fall. This correlation between tephra and pollen represents an important aspect of tephrochronology, that it can provide a dating framework against which other dating techniques can be calibrated and validated.

Dendrochronology and other growth-ring studies

Living things respond to time passing and in doing so they may show signs of incremental growth. In shelled animals such as scallops these growth markers are clearly seen as concentric patterns on the outside of the shell (Fig. 8.10).

The calcareous shells of modern corals also show evidence of incremental growth that is influenced by factors such as water temperature, water quality and the availability of sunlight. The coral grows because the animal deposits calcium carbonate $(CaCO_3)$ every day of its life and these daily layers are visible under a microscope. Species of coral living in temperate oceans are subjected to seasonal variations in temperature so winter growth differs from summer growth and it is thus possible to recognize yearly growth lines on the coral skeleton (Fig. 8.11).

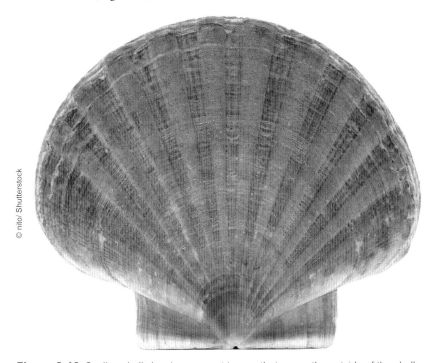

© nito/ Shutterstock

Figure 8.10 Scallop shell showing concentric growth rings on the outside of the shell.

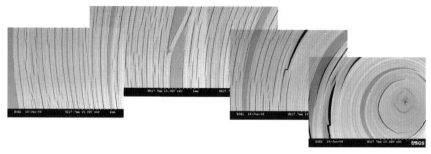

Figure 8.11 Scanning electron microscope photograph of growth rings in black coral from the Gulf of Mexico. USGS image.

Figure 8.12 Fossilized horn coral (Heliophyllum) of Devonian age, showing growth rings. USGS image.

Heliophyllum is a variety of extinct coral found as fossils in rocks of Devonian age (approximately 416–360 mya). *It* was a solitary coral rather than a reef-building colonial variety. Its distinctive laminated form (Fig. 8.12) is clearly periodic, its growth being a function of time and, depending on how well the fossil is preserved, it may be possible to differentiate not only annual cycles of growth but also lunar-influenced cycles and even diurnal or daily patterns of growth. The patterns have been employed to calculate the length of the Devonian day and year.

Fossil corals like *Heliophyllum*, living in the Devonian Period display years of about 400 lines, equating to 400 days, showing that the Earth did indeed turn faster at that time. Evidence from other fossil groups, including brachiopods and nautiloids confirm that the number of days per year in the Devonian would have been around 400, since a year, the time taken for the Earth to orbit the Sun, has not changed appreciably.

For corals that lived during the upper Carboniferous (300 mya), there are approximately 380 lines each year. The fossil record thus clearly demonstrates that Earth's rotation has gradually slowed over time, and that it is still slowing down today. The actual length of the day is increasing and has been doing so for billions of years. The Moon's gravitational pull causes the Earth's rotation to slow down and so the length of a day, which is the time taken for the Earth to revolve on its own axis, has lengthened. The tidal braking in the Earth's rotation is due primarily to friction in the oceans, for example bottom friction induced by tidal currents flowing across the seabed or by various types of breaking waves.

This slowdown is exceedingly small, currently estimated at about 2–3 milliseconds per day per century, but going back several hundred million years this makes an appreciable difference to the length of the day (around 2–3 hours shorter) and therefore the overall number of days in a year would be greater.

Most people are probably aware that the growth stages of a tree are measured in the concentric rings seen across the trunk, where each ring represents a yearly increment (Fig. 8.13). The thickness and condition of these rings is influenced by environmental factors such as the availability of moisture and sunlight so that as well as measuring the passage of time the trees also record past climate changes.

Figure 8.13 Annual tree growth rings. © Don Pablo/ Shutterstock

The information on past environmental conditions and the passage of time from the study of fossil corals or the measurement of tree rings are just two techniques in what can be described as 'climate archives'. Other methods of investigating past environments include the examination of ice cores, the identification of pollen and spores within peat bogs or lake sediments, and the growth of stalagmites and stalactites. Ice cores such as those taken from the Antarctic ice cap can give us information on the carbon dioxide (CO_2) and methane (CH_4) content of the atmosphere over the last several hundred thousand years, very relevant in our attempts to understand global warming. The growth of stalagmites and stalactites are also useful temperature indicators and the pollen and spore content of lake sediments allows a reconstruction of prehistoric landscapes and environments. However, the development of dendrochronology in particular has had a major impact on the dating of events and artefacts in archaeology and the development of an accurate, worldwide, calibrated chronology based on tree rings has been one of the great success stories in modern science.

Development of dendrochronology

Bristlecone pines are found at high altitudes in the arid regions of the south-western states of the USA (Fig. 8.14). They are the oldest known individuals of any species. It is thought some examples are more than 5,000 years old and they have played a vital role in the development of the science of dendrochronology. Their wood is extremely dense and resinous and so is resistant to attack by insects and fungi and other pests, accounting for the remarkable durability of the trees. They are slow growing, seldom exceeding 20 m in height and tending to be broad and squat in shape. They generally retain their needles for up to 30 years, which helps conserve valuable energy. A noticeable feature of bristlecone pines is that they remain rooted and standing for centuries after their death as the cold, dry climate inhibits decay.

They were first utilized in chronology in the early years of the 20th century by A. E. Douglass, an astronomer at the University of Arizona. He used tree rings to investigate sunspot activity. He reasoned that changes in solar output would cause climate change on Earth which would result in variations in the growth rings of the trees. From these observations the Arizona tree-ring record revealed an 11-year solar cycle of sunspot activity. Douglass went on to expand the tree-ring record in Arizona by hundreds of years by overlapping the growth-ring patterns of living trees with the patterns identified in timbers taken from archaeological sites. By the early 1920s Douglass had constructed a chronology that extended back to 1284.

Figure 8.15 shows how cross-dating of tree rings works. Samples taken from living trees (a) and dead wood, either in the field (b) or as part of a structure

Figure 8.14 Bristlecone pine, south-western USA. © KennStilger47/ Shutterstock

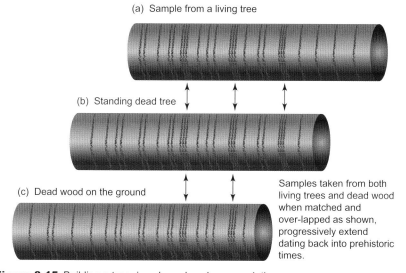

(a) Sample from a living tree

(b) Standing dead tree

(c) Dead wood on the ground

Samples taken from both living trees and dead wood when matched and over-lapped as shown, progressively extend dating back into prehistoric times.

Figure 8.15 Building a tree-ring chronology by cross-dating.

(c), when matched and overlapped as shown can progressively extend dating back into prehistoric times. This chronology can be used as a method for dating events such as a volcanic eruption or a change in climate that caused a marked change in the appearance of a ring or a set of rings. If the tree has been used in a construction the date of felling can be calculated, thus allowing the building to be dated.

151

Factors other than solar activity can cause variations in the tree-ring pattern. Many trees in temperate regions make a single growth ring each year, growing from the centre out, and this ring reflects the climatic conditions for that year. The widest rings occur when sunshine and rainfall conditions are favourable, while drought conditions can result in abnormally narrow rings. The eruptions of the Laki volcano in Iceland in 1783 and Mount Tambora in Indonesia in 1815, mentioned in the earlier section on tephrochronology, were responsible for widespread cold conditions in the northern hemisphere in the summers immediately following the eruptions. This resulted in markedly narrow growth rings in oak trees in Ireland. In the case of the Laki event there is a cross-reference between the tree-ring characteristics and the Laki ash layer identified in the sedimentary record in Ireland and Scotland.

Sampling and dating

Rather than cut down the whole tree to enable the tree rings to be counted it is possible to extract a small-diameter core using an appropriate drill. Douglass' data on beams from archaeological sites in Arizona have been extensively used in many other parts of the world, with particular success in Ireland. Here, settlement of the landscape had taken place following the retreat of the last ice around 10,000 years ago, with the earliest settlements (Mesolithic) probably occurring around 9,000 years ago. Oak timbers were available for all periods of the archaeological record, from living trees, historical buildings, archaeological sites and bog oak. The chronology currently extends back beyond 7,000 years ago. The ability to achieve such remarkably precise dates in archaeology represents an enormous advance in interpretation in the last few decades and has transformed our understanding of many sites around the world. Two examples serve to illustrate this:

- Emain Macha (or Navan Fort), in County Armagh, Northern Ireland, is a huge circular earthwork enclosing 5 hectares on a raised site. There is archaeological evidence that it once housed a great wooden structure and is likely to have been used for rituals, probably with aristocratic connections. Professor Mike Baillie and his colleagues at Queen's University, Belfast, have obtained a date from dendrochronology for a large central oak post in the building which last grew in 95 BCE and was probably felled either late in 95 or early in 94 BCE.
- In the 14th century, around 1347, bubonic plague, known as the Black Death, arrived in Europe, spreading rapidly and devastating whole populations with the deaths of millions of people. Dendrochronologists found it virtually impossible to bridge the gap in chronologies based

on wood from buildings. It seems that there was a hiatus in building that coincided with the worst period of the plague epidemic. When populations are suffering devastating losses there is less priority given to new building and this phase lasted for nearly 100 years. It is interesting that this reduction in population associated with the epidemic led to a regeneration phase for oak forests, seen in the tree-ring record. This regeneration possibly occurred because previously managed woodlands had been abandoned through a lack of forestry workers.

So, as well as being a powerful tool for absolute dating for archaeological and historical events, dendrochronology clearly has an important environmental dimension. Tree rings have much to tell us about such things as climate changes, volcanic eruptions, meteorite impacts and social changes. The resolution of tree-ring data is one year, among the highest resolutions available, but with a total range of around 9 kyr, dendrochronology is useful over a relatively short period. However, one of the other important roles of tree ring data has been as an invaluable calibration mechanism for the radiometric timescale developed for archaeology, based on the isotopes of the element carbon (C), our next consideration.

Radiocarbon dating

Radiometric dating techniques for geological processes were discussed in an earlier chapter and carbon-14 (^{14}C) dating is based on the same principles. The radioactive decay reactions used to measure the age of geological processes generally have very long half-life periods, the half-life being the time taken for half of the parent atoms to decay to daughter atoms (see Chapter 7). The uranium–lead decay series, for example, has a half-life of 4.5 byr. It is necessary to use series with long half-lives because of the very great ages of many of the geological processes being dated, but they are not suitable for archaeology. Here the ages are of the order of thousands of years rather than millions. The most widely used radiometric dating technique in archaeology is radiocarbon dating using the isotopes of carbon. The method was developed in the 1940s and early 1950s by Willard F. Libby and his team at the University of Chicago and it has become the most powerful method of dating artefacts and events up to around 50 kyr old.

The Earth's atmosphere contains various isotopes of carbon in roughly constant proportions. These include the main stable isotope (^{12}C) and an unstable isotope (^{14}C). The ^{14}C isotope is formed when as a result of cosmic radiation from the Sun nitrogen (N) atoms are changed into ^{14}C atoms which are radioactive and which decay to give nitrogen ^{14}N atoms (Fig. 8.16). Plants absorb both forms from carbon dioxide (CO_2) via photosynthesis in the atmosphere.

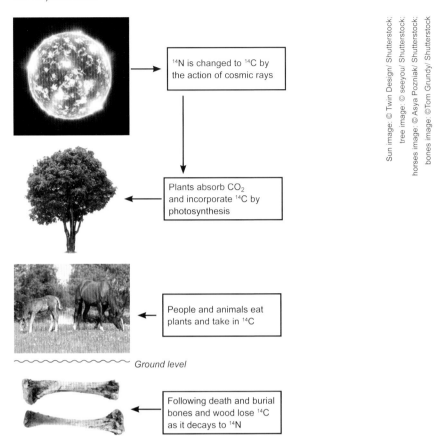

Figure 8.16 Showing the pathway from ^{14}N to ^{12}C and back to ^{14}N in the natural cycle.

When an organism dies, it contains the standard ratio of ^{14}C to ^{12}C, but as the ^{14}C steadily decays the proportion of ^{14}C decreases at a known constant rate. This rate is the time taken for it to lose half of its ^{14}C atoms, the half-life of ^{14}C, which is 5,730 years. Measuring the remaining proportion of ^{14}C in organic matter will give an estimate of its age and dates derived from carbon samples can be calculated back to around 50,000 years.

Calibration

Radiocarbon ages are always reported in terms of years BP (before present), where the present is taken as the year 1950. There are a number of reasons why radiocarbon measurements are not true calendar ages. The age is calculated on the proportion of radiocarbon found in the sample and it is assumed that the atmospheric radiocarbon has always been the same as it was in 1950 and that the half-life of carbon is 5,568 years, the original value measured by Willard Libby. Although less accurate, the Libby value was retained to avoid inconsistencies when comparing ^{14}C test results produced originally. This discrepancy between the half-life period used in the calculation and the true figure of 5,730 years, plus the fact that the proportion of radiocarbon in the atmosphere has varied by a few percentage points over time means that radiocarbon measurements need to be corrected. The small fluctuations in the ratio of ^{14}C to ^{12}C in the atmosphere over time, as recorded in sequences of stalactites in cave deposits and in tree rings, allow the adjustment or calibration of the unadjusted radiocarbon age, to give a more accurate estimate of the calendar date of the material. This is now usually done using tree rings as a radiocarbon record.

In principle this is very simple. To calibrate radiocarbon determinations the proportion of radiocarbon in the sample is matched with a tree ring with the same proportion. A typical example of a calibration curve is shown below (Fig. 8.17). The vertical axis shows the radiocarbon concentration expressed in years BP. The horizontal axis shows calendar years derived from tree-ring data.

The correlation between the unadjusted ^{14}C determination and the tree-ring date can then be read off from the graph. Since there are precision limits on the analysis of both the tree rings and the sample then clearly there will always be a range of possible calendar years.

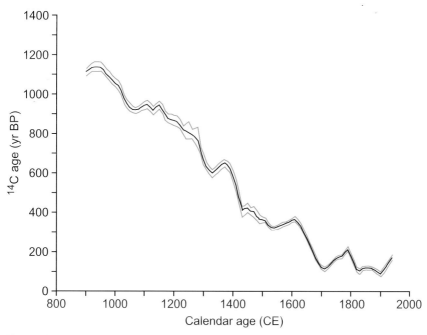

Figure 8.17 Plots of true age (from tree rings) against radiocarbon age. After Stuiver and Pearson.

While radiocarbon determinations and dendrochronology are not the only scientific dating methods available to archaeologists, in many areas of the world, for example Ireland, they are the most widely used chronological methods. The calibration curves for ^{14}C measurements using tree-ring data have been replicated in high-precision laboratories in several countries, using a range of tree species, including oak in northern Europe and bristlecone pine in the USA. The addition of important environmental information to an accurate absolute timescale for the last several tens of thousands of years has been of major benefit, not just to archaeologists but to a wide spectrum of scientists studying palaeo-environments, studies particularly relevant today given the changes the contemporary environment is undergoing.

The writer L. P. Hartley said 'The past is a foreign country: they do things differently there'. Rather than the present being the key to the past as Lyell and the uniformitarianists would have it, the key to the present might well be in the past. The fuller and more precise the data we have from that 'foreign country' the better for the welfare of future generations.

In previous sections we have investigated the passage of time using a variety of stratigraphical methods – lithostratigraphy, based on the physical characteristics of the rocks, and biostratigraphy based on the fossils contained in the rocks. We now consider the measurement of time using the secular variation of one of the Earth's fundamental properties, its magnetic field, the basis of magnetostratigraphy and magnetochronology.

Magnetochronology/magnetostratigraphy/ archaeometric dating

Figure 8.18 is a view of the night sky in the northern hemisphere showing the aurora borealis or Northern Lights, a spectacular green and purple glow at altitudes of 100 km or more, powered by energetic particles at the edge of space. The Northern Lights are probably the most spectacular manifestation of the Earth's magnetic field. They are the result of the interaction of the solar wind – a stream of charged particles (electrons and protons) escaping the Sun – and the planet's magnetic field and atmosphere. The solar wind distorts the Earth's magnetic field, the geomagnetic field, and allows some of those particles from the Sun through the geomagnetic field at the magnetic north and south poles. Here the atmospheric gases are excited by the charged particles and they glow in the same way as the gas in a fluorescent tube, thus creating the phenomenon we see here. The variety of colours is due to the type of gas particles that are in collision. The most common colour in the aurora is a pale yellowish-green, produced by oxygen molecules located nearly 100 km above the Earth. Nitrogen produces a blue or purplish-red aurora.

Figure 8.18 The aurora borealis, or Northern Lights. © SurangaSL/ Shutterstock

The aspect of the geomagnetic field familiar to most people is its ability to influence the movement of a compass needle and indicate the direction of magnetic north. The naturally occurring mineral magnetite, a common oxide of iron often called lodestone, is attracted to a magnet and may also act as a magnet and attract iron particles. The ancient Chinese noticed that a suspended or floating piece of magnetite aligned itself consistently with the North Magnetic Pole and this led eventually to its use as a compass for navigation. The lodestone spoon device shown in Figure 8.19 was initially used by the Chinese in astrological and divining ceremonies before eventually being adapted for navigational purposes and as long as it can move freely it will always come to rest in an approximately north–south direction.

A further application of the Earth's magnetic field is its use by certain groups of animals as a navigational aid. Homing pigeons, for example, use it to find their way home over long distances. It seems that the birds have tiny particles of magnetic iron ore (magnetite) in their beaks and these are used to sense the orientation of the geomagnetic field and to use this information as a navigational aid. A number of animal groups, including insects, turtles and sharks as well as birds, appear to have this ability to detect a magnetic field and use it to perceive direction, altitude or location and develop regional maps.

Figure 8.19 Chinese compass from the 4th century BCE. © hjochen/ Shutterstock

A compass functions as a pointer to 'magnetic north' because the magnetized needle aligns itself with the lines of the geomagnetic field. The magnetic field pulls one end or pole of the needle towards the Earth's North Magnetic Pole, and the other towards the South Magnetic Pole.

If we plotted the direction in which a compass needle comes to rest at various points in space around the Earth the pattern would look like Figure 8.20. It is roughly symmetrical about the Earth's centre which suggests the origin of the geomagnetic field is inside the Earth. The field is described as a magnetic dipole; in other words it has two poles and the pattern is similar to that of a simple bar magnet.

What causes the Earth's magnetic field?

The geomagnetic field is believed to originate from the Earth's core. Using the passage of earthquake waves through the Earth, geophysicists have been able to construct a likely model for its internal structure (Fig. 8.21).

The outer layers of the Earth are the crust and mantle. As described earlier (Chapter 5) the mantle is a plastic solid capable of sustaining convection currents which provide the driving force for plate tectonics. Beneath

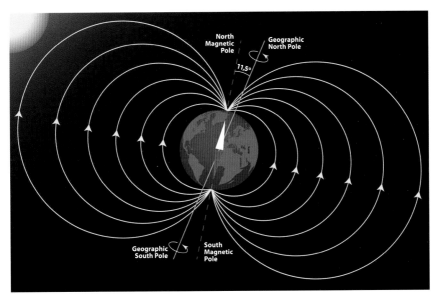

Figure 8.20 The Earth's magnetic field. © Milagli / Shutterstock

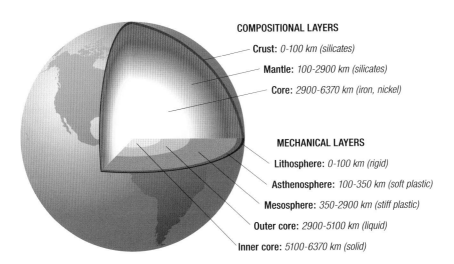

Figure 8.21 The inner structure of the Earth showing the main divisions: crust, mantle and core. © Peter Hermes/ Shutterstock

the mantle is the core, which is about two-thirds the size of the Moon and composed primarily of nickel–iron. It consists of an inner core and an outer core. The inner core is solid and at a temperature of around 5,700°C. The outer core is a 2,000-km-thick zone which is liquid. It is this outer, liquid core which generates the geomagnetic field.

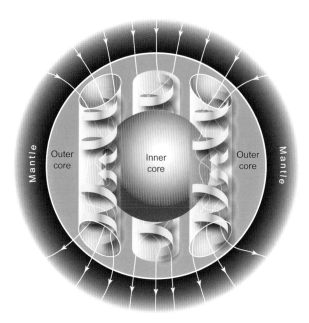

Figure 8.22 View of the Earth's solid inner core and the outer core with the helical convection cells which create circulating electrical currents which generate the magnetic field. After USGS.

Within the outer core differences in temperature, pressure and composition form convection currents in the liquid metal and this effect is increased by the forces caused by the Earth's rotation to form spiral or helical currents (Fig. 8.22). While the origin of the geomagnetic field is not completely understood, it is thought to be associated with electrical currents produced by the coupling of these convective effects and rotation in the spinning liquid metallic outer core of iron and nickel.

This mechanism is termed the dynamo effect – the flow of liquid iron generates electric currents, which in turn produce magnetic fields. A dynamo is a machine for converting mechanical energy into electrical energy and a simple form is commonly used on bicycles to generate power for a lamp. In the bicycle dynamo the magnet is turned inside a wire coil by the rotation of the wheel. This creates a changing magnetic field which generates electricity, thus lighting the lamp bulb.

In summary, the motion of the electrically conducting iron in the presence of the Earth's magnetic field induces electric currents. Those electric currents generate their own magnetic field and, as the result of this internal feedback, the process is self-sustaining so long as there is an energy source sufficient to maintain convection. This self-sustaining or self-exciting loop is known as the geodynamo.

The relatively large, hot core and the rapid spin of the Earth probably account for the exceptional strength of its magnetic field compared with those of the other terrestrial planets. Venus, which has a metallic core that may be similar to Earth's in size, rotates very slowly and has no detected internal magnetic field, while Mercury and Mars have only small intrinsic magnetic fields.

Polarity reversals

An important feature of the Earth's geomagnetic field is polarity reversal – where the direction of the magnetic field reverses direction periodically and the North Magnetic Pole becomes the South Magnetic Pole, and vice versa. By examining the direction of magnetism of many rocks it is known that such reversals occur at intervals varying from tens of thousands of years to millions of years. Certain rocks, particularly basalt lavas, can retain within their component minerals, particularly magnetite, the characteristics of the geomagnetic field that existed at the time of eruption of the lava. This information is known as the remanent magnetization and gives among other features the polarity of the geomagnetic field at the time.

In igneous rocks the remanent magnetization is described as thermoremanent magnetization. It is acquired as the temperature of the igneous rock falls below a critical level and the magnetite crystals in it align themselves according to the characteristics of the geomagnetic field in which the rock is located, thus making a permanent record of its orientation. It is also possible for sedimentary rocks to have remanent magnetization – in this case depositional remanent magnetization which is acquired by the rock when magnetic grains in the rock may align themselves with the current magnetic field during or soon after deposition. This means that in a sequence of basalt lavas or sediments with magnetic grains the direction of the remanent magnetic polarity can be used as the basis for the subdivision of the sequence into units characterized by their magnetic polarity.

Using either of these forms of remanent magnetization is directly analogous to the use of the fossil content of rocks to set up the stratigraphical column. The study of the remanent magnetic characteristics of rocks is called palaeomagnetism, and using radiometric dates a Geomagnetic Polarity Time Scale (GPTS) has been constructed. This is a record of the many episodes of the reversal of the Earth's magnetic field.

The GPTS was developed by accurate age-dating of rocks from around the world, during which it was noticed that rocks from particular time periods contained minerals whose magnetic orientation was opposite to the field currently in operation. Comparison of a pattern of magnetic reversals from a sequence of rocks with that of a sequence of known age allows the approximate age of the unknown sequence to be established.

Figure 8.23 shows the Earth's Phanerozoic geomagnetic polarity timescale with its various subdivisions known as chrons which are normal and reversed intervals approximately 1myr to 10 myr. Sub-chrons are short reversals within larger chrons, representing intervals of approximately 100kyr to 1myr. If the palaeomagnetic field had a similar orientation to the present-day field, that

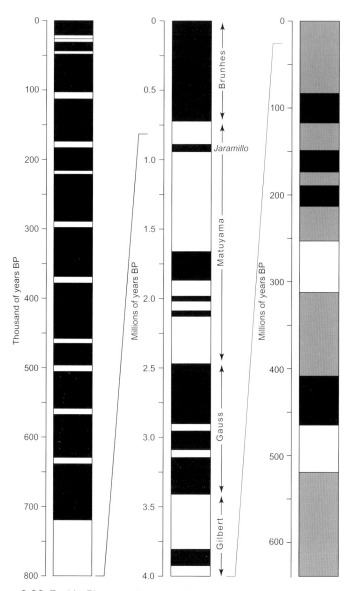

Figure 8.23 Earth's Phanerozoic magnetic polarity timescale. Black areas represent periods of 'normal polarity' (corresponding to the present-day polarity); white areas are periods of reversed polarity; grey areas on the right indicate periods of mixed polarity interspersed between extended periods (superchrons) of constant polarity. After Vita-Finzi & Fortes.

is, with the North Magnetic Pole close to the North Geographical Pole, then the field is described as having normal polarity, shown on the diagram below as black. Conversely, if the field has an orientation so that the North Magnetic Pole was near the South Geographical Pole, then it is described as having reversed polarity, shown on the diagram as white.

These magnetic reversal sequences are given in the column to the left. This has been a useful tool in the understanding of the sequence of events in the major basalt plateaus of the world such as the Columbia River Plateau in the western USA, but it has been particularly useful for studying the basalts of the oceanic crust which solidify from molten lava symmetrically about the mid-ocean ridges in the process of sea-floor spreading in plate tectonics. The Late Jurassic (160-150 mya) is the age of the oldest existing oceanic crust.

Magnetostratigraphy and sea-floor spreading

The plate tectonic cycle begins by continents splitting apart to form a new ocean; for example the Red Sea is widening as Arabia moves away from Africa and will eventually become an extension of the Indian Ocean. This process of widening is called sea-floor spreading and is responsible for creating new crust as the ocean develops. After the Second World War, as has often been the case in the past, technologies developed for some aspect of warfare were adapted for peace-time use after the end of hostilities. In the 1950s scientists began using airborne instruments which had been developed to detect enemy submarines for measuring the magnetic field (magnetometers). They began to recognize variations in the magnetic field of the ocean floor. This phenomenon had been known to Icelandic sailors as long ago as the late 18th century when they were aware of the possibility of distorted compass readings in certain areas. The degree of variation is due to the magnetic mineral magnetite, an important constituent of basalt which makes up much of the ocean floor.

A great deal of the evidence confirming this process of ocean-widening by gradual addition of lavas erupting on the seabed came from the recognition of polarity reversal in the remanent magnetism of these oceanic basalts. As magma is erupted onto the seabed it cools and assumes the polarity characteristic of that time, normal or reversed. Subsequent eruptions after a polarity change will therefore show the opposite polarity, so that a series of symmetrical magnetic stripes can be detected on the sea floor as the spreading takes place in both directions away from the ridge. This allows the sea-floor profile to be matched to the magnetic polarity timescale and so the various stages of sea-floor spreading can be dated. Figure 8.24 shows the development of a symmetrical pattern of anomalies on either side of the mid-ocean ridge for the last 5 myr. New oceanic crust forming continuously at the crest of the mid-ocean

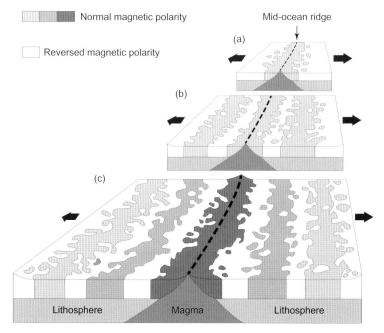

Figure 8.24 A theoretical model of the formation of magnetic striping. USGS image.

ridge cools and becomes increasingly older as it moves away from the ridge crest during sea-floor spreading. The labels are: a) the spreading ridge about 5 mya; b) about 2–3 mya and c) present day.

Magnetostratigraphy therefore uses the polarity reversal history of the geomagnetic field recorded in igneous or sedimentary rocks to determine the age of those rocks. A major advance in recent years has been the application of this timescale, developed on the sea floor, to the fields of archaeology and anthropology, and in particular the study of the origins of our own species, *Homo sapiens*.

Magnetostratigraphy in archaeology and anthropology

In the field of anthropology, defined as the study of human beings, the dating of sediments and volcanic rocks associated with early hominid sites during the last 5 myr or so has been greatly improved by the use of magnetostratigraphy. The term hominid in this context refers to humans and the relatives of humans. The field of science which investigates early human development and evolution is known as palaeoanthropology – the meeting of palaeontology and anthropology.

By measuring the polarity characteristics of a sequence of rocks it is possible to correlate a local polarity sequence with the Geomagnetic Polarity Time Scale, which is calibrated in millions of years. The Geomagnetic Polarity Time Scale (GPTS) has been constructed from an analysis of magnetic anomalies measured over the ocean basins linked to known and dated magnetic polarity reversals found on land.

On the GPTS for the late Cenozoic, the Brunhes–Matayuma Reversal (see Fig. 8.25) was the last major reversal of the Earth's magnetic field. This occurred approximately 780 kya. Since early human forms began to appear around 4 mya, the Brunhes–Matayuma Reversal is a significant marker horizon in the field of palaeoanthropology.

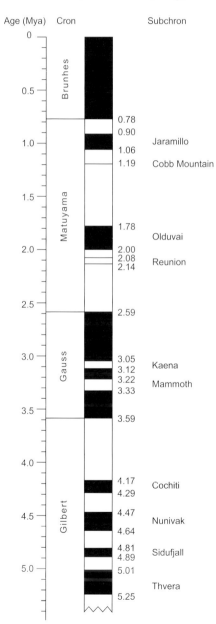

Figure 8.25 The Geomagnetic Polarity Time Scale (GPTS) during the last 5 myr. Normal polarity is shown as black, reversed polarity as white. After USGS image.

Figure 8.26 Olduvai Gorge, Tanzania. © Alexandxer Kuguchin/ Shutterstock

The Olduvai Gorge (Fig. 8.26) is an important palaeoanthropological site in northern Tanzania in East Africa. It is a steep-sided ravine consisting of two branches that have a combined length of about 48 km and are 90 m deep. Deposits exposed in the sides of the gorge cover a time span from around 2 mya to 15 kya. The deposits were laid down across a number of polarity events, allowing the beds and associated archaeological finds to be dated using magnetostratigraphy. The gorge has yielded the fossil remains of more than 60 members of the human lineage, providing the most continuous known record of human evolution during the past 2 myr, as well as the longest known archaeological record of the development of stone-tool industries.

The Olduvai subchron (a short-term normal period within a longer mainly reversed period, see Fig. 8.25) was named after its discovery locally at the Olduvai Gorge. It was one of the first magnetic polarity subchrons to be identified and occurred between 1.78 and 2.0 mya.

Similar geomagnetic dating in China has allowed the dating of hominid remains with the recognition of short reversal events within the Brunhes chron at around 0.7 myr.

Measurement of the reversals of the geomagnetic field and the construction of a polarity time scale has proved to be a powerful tool not only in the dating of sea-floor spreading processes but also in charting the evolution of our own ancestors from the earliest hominid forms, over the last few million years. However, geomagnetism has more to offer dating than just field polarity reversal – the intensity or strength of the magnetic field and the position of the geomagnetic pole relative to the geographical pole also have a role in the dating of archaeological events.

Polar wandering and archaeomagnetic dating

As well as being capable of changing orientation from normal magnetization to reverse magnetization, the geomagnetic pole can also change its position in relation to the geographic pole. When using a simple compass the needle points to geomagnetic north and not geographic north. The north geomagnetic pole is currently in northern Canada and the map below (Fig. 8.27) shows how its position has changed over the recent and historical past. The geomagnetic field is produced by currents in the liquid iron that makes up the Earth's outer core and so is the result of a dynamic and changing process. These changes in the geomagnetic field are relatively rapid compared with most of the geological processes we have been considering. For example, the angle

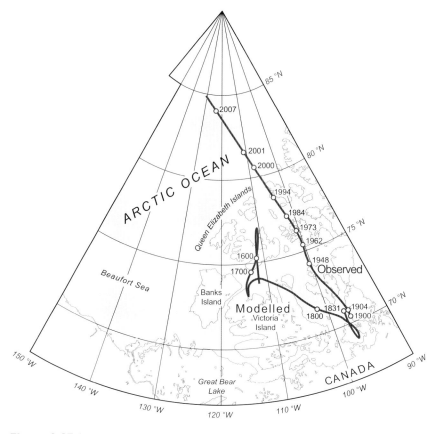

Figure 8.27 Magnetic north pole positions of the Earth for the recent and historical past. The pole is defined as the position where the direction of the magnetic field is vertical. After Vita-Finzi & Fortes.

between geographical north and geomagnetic north where I am currently writing (County Down in Northern Ireland) is 3 degrees 43 minutes West and changing by a few minutes each year, but if I were writing in London the angle would be 1 degree 5 minutes West, and also changing by a few minutes each year.

Figure 8.27 shows how the North Magnetic Pole has shifted since 1600. Notice that the distance between yearly points has increased markedly since the middle of the last century and is now moving about 40 km per year. Joining the points showing the pole positions traces the polar wandering curve since 1600. From this it can be seen that the present north-westerly path started between 1800 and 1900. The axis of the Earth's magnetic field is inclined at roughly 11 degrees to the rotational axis. During the last approximately 7 kyr the north geomagnetic pole has always been within about 11 degrees of the geographical pole, and so for the historical past the geomagnetic field has averaged around the geographical pole.

Archaeomagnetic dating measures this wandering of the North Magnetic Pole and can utilize both direction and strength of the geomagnetic field. This is because it is not only the position of the magnetic poles that changes but also the intensity of the magnetic field. The field intensity determines the degree of attraction of a compass needle to the magnetic poles and the strength of magnetization acquired by magnetic minerals. Over the last 150 years observations in London and Paris indicate that while the Earth's field has changed in direction, drifting westwards by about 0.25 of a degree of longitude each year, the intensity or strength of the field has changed by about 0.05% each year. These long-term changes in the terrestrial magnetic field, known as secular variations, form the basis for the archaeomagnetic dating technique.

Just as remanent magnetism forms the basis of magnetostratigraphy as described earlier, so also archaeomagnetism depends on it. Most archaeological materials include magnetic particles, if only as impurities. If these have been heated above a specific temperature for that magnetic mineral (generally around 600°C) and then allowed to cool, they will retain a magnetism that will give the direction and intensity of the geomagnetic field at the time of cooling. This is thermoremanent magnetism (TRM) and is the same phenomenon exhibited by basalt lavas as they cool. The magnetic particles now point to the location of magnetic north at the time of heating. Fired clay materials such as pottery, bricks or tiles can therefore be dated this way, along with stationary features such as hearths, fireplaces and kilns where the firing took place. The TRM measured is the last time the feature was fired; if it is reheated to a similar temperature a new magnetization will be imposed.

By using another dating method such as dendrochronology or radiocarbon dating to obtain the absolute age of an archaeological feature such as a hearth, and measuring the direction of the magnetism in the hearth clay today, it is possible to determine the location of the magnetic north pole the last time the hearth was fired. This process can be carried out if the direction of the current magnetic north pole can be marked on the hearth and a magnetometer is then used to measure the orientation of the iron particles in the hearth. This gives the virtual geomagnetic pole or VGP. Using a large number of these VGPs a curve showing polar wandering with time can be constructed. This is used as a record against which VGPs of unknown age can be compared and assigned a date.

Dating based on variation in the geomagnetic field intensity has the advantage that unoriented material can be used. The use of intensity variation means that, for example, fired clay pottery fragments commonly found on archaeological sites can provide valuable information.

The graph in Figure 8.28 shows variation in the strength of the dipole field over the last 10 kyr for data obtained from archaeological hearths in Europe and serves to illustrate the main difference in the measurement of time in archaeology compared with geology. Apart from the dating used by palaeoanthropologists investigating early hominid evolution, the events and artefacts being dated in archaeology tend to be of the order of thousands of years old, rather than the millions or billions of years required to date geological events or processes.

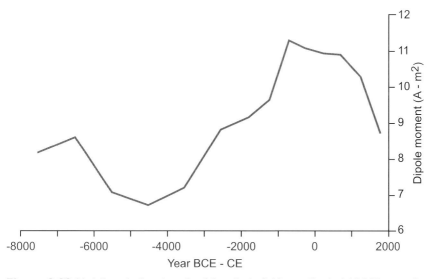

Figure 8.28 Variations in the strength of the dipole field over the last 10,000 years for archaeological hearths in Europe. After McElhinny and Senanayake.

Figure 8.29 Replica of the Laetoli footprints left by early hominids on a layer of freshly deposited ash in Tanzania, nearly 4 mya. (Courtesy Smithsonian's Human Origins Program.)

Despite this discrepancy there has been remarkable overlap in techniques for the measurement of time between geology and archaeology. Perhaps one of the best examples of this symbiotic relationship has been the spectacular footprints found at Laetoli, Tanzania, not far from the Olduvai Gorge (Fig. 8.29).

These footprints are part of a 24 m long line of impressions in freshly deposited volcanic ash and provided convincing evidence of bipedalism, or walking on two legs, in early hominids in the Pliocene Period. Using a combination of stratigraphy and radiometric dating the footprints were dated at 3.8–3.6 myr. Like the casts of the bodies buried in the Fugitives Garden at Pompeii by the ash flow at Vesuvius (see Fig. 8.5), this hominid trackway made by two or possibly three of our early ancestors gives us an insight into an event that happened a long time ago. They may have been a family group and it is impossible not to speculate on their eventual fate since they were obviously living in close proximity to an active volcano.

The discussion of time in archaeology and the various techniques used to measure it has brought the narrative up to the historical past and to the present and the next consideration is time in the future. In terms of the stratigraphical column the current epoch is the Holocene. There are suggestions, however, in the geological world that we are already beyond this and are entering the Anthropocene – the most recent geological period and the one where the Earth's population, for the first time, is the greatest influence on the future of the planet. It is time to think about future time.

Chapter 9

Time future

'The future depends on what you do today...'
Mahatma Gandhi (1869–1948)

Into the Anthropocene

The distribution of lights on the Earth's surface as seen from space (Fig. 9.1) gives some idea of the extent of human activities and our influence on global ecosystems.

The scale of human activities on Earth is such that it is capable of disrupting or changing the natural cycles we depend on for our existence on the planet. The rapid rise in the world population and the attendant rise in resource consumption are now influencing such things as the exchange of water between land and ocean, the movement of water in the oceans and the composition of the atmosphere.

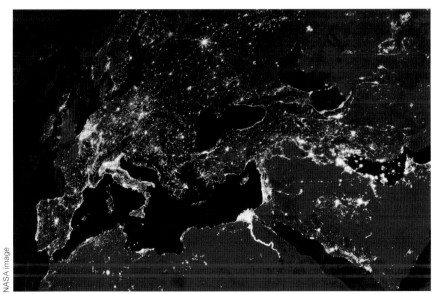

NASA image

Figure 9.1 Earth night view from space with city lights over Europe and the Middle East..

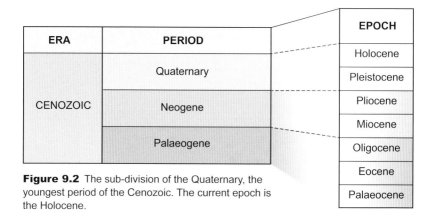

ERA	PERIOD	EPOCH
		Holocene
CENOZOIC	Quaternary	Pleistocene
	Neogene	Pliocene
		Miocene
	Palaeogene	Oligocene
		Eocene
		Palaeocene

Figure 9.2 The sub-division of the Quaternary, the youngest period of the Cenozoic. The current epoch is the Holocene.

In our examination of how time is measured in geology we have looked at the construction of the stratigraphical column and how it reflects a combination of relative time and absolute time. The youngest era of this stratigraphic column, the Cenozoic, is shown in Figure 9.2.

The final period of the Cenozoic is the Quaternary and the most recent epoch of this period has been designated the Holocene, beginning around 10 kya with the ending of the last ice age and lasting to the present. The name Holocene is derived from the Greek words *holos*, meaning whole or entire, and *kainos*, new, together meaning 'entirely recent'. The Holocene also includes the development and impact of the human species worldwide, the Cognitive Revolution, including the rise of major civilizations, the appearance of written histories and the Agricultural Revolution, and the accelerated transition to urban living following the Scientific and Industrial Revolutions.

While all of these changes have taken place entirely in the epoch currently designated as the Holocene, there is increasingly a view in scientific circles that the Earth has entered a new phase, one that is sufficiently different to be a new epoch and to warrant a new name – the Anthropocene. The term derives from the Greek word *anthropos* meaning humans in general and referring to the role of the Earth's population in the major changes seen in the environment – the view that humans represent the principal geological agent on the planet. The term is still informal, in the sense that it has not yet been officially adopted as a formal unit in stratigraphy, but the process of considering whether it should be formally accepted into the geological timescale has already begun. Among scientists there is considerable debate about when the Anthropocene started – was it the beginning of farming, the start of the Industrial Revolution, the dawn of the Atomic Age or the period in the second half of the 20th century referred to as the Great Acceleration when populations and resource consumption rose dramatically all over the world?

The people of the Neolithic, around 10 kya, began to develop agriculture and were thus the first people to significantly alter their environment. They began deforestation and clearing land for cultivation and in the process developed new and better tools for planting and harvesting (see Figure 2.5). Since agriculture ties populations to the land, people began to live in permanent settlements that grew eventually into the first cities. Many of the problems affecting the environment today, such as deforestation, soil erosion and water shortages associated with excessive irrigation, began with the earliest farmers. The change from stone to bronze to iron for farming implements also meant more efficient weapons were available, an early form of the arms race which we can also recognize today.

Another possibility for the start of the Anthropocene is the beginning of the Industrial Revolution in the late 18th century in Britain and spreading to other parts of the world in the following 100 years. Arguably the Industrial Revolution was the most profound social upheaval in history since the development of agriculture in Neolithic times. Never before had the lives of ordinary people changed so much, at least not until the technological changes around the end of the 20th century. The harnessing of steam power fuelled by coal led to the development of industrial-based societies, a consumer driven economy and the beginning of the urbanization trend that continues to this day. The Industrial Revolution brought benefits and costs. Increased general prosperity has to be balanced against the health costs of coal burning and the toxic materials used in many of the industrial processes. However, it laid the foundation of the globalized economy most of us live in today, even though problems such as poor sanitation and the lack of a clean water supply still blight far too many of the planet's population.

The steady increase in urban living which has occurred worldwide since the beginning of the Industrial Revolution means that there have been major changes on the face of the Earth. The pattern of lights across the whole Earth as seen from space (see Fig. 9.3) clearly shows the position of the great metropolitan areas in Britain and the rest of Europe, the eastern seaboard of North America and the densely populated urban regions of India and China. These brightly patterned areas are in stark contrast to the central regions of the South American and African continents. In Africa lights are mostly visible along the Mediterranean coast and the Nile valley in the north, and around South Africa, while in South America the vast emptiness of the Amazon Basin is marked by a lack of illumination visible from space.

In our earlier examination of the fossil record and the stratigraphical column it was clear that the major subdivisions or periods were marked by some major event, often a major change in climatic conditions or a major extinction period. Although the time human-like species have been on Earth is very short, a few

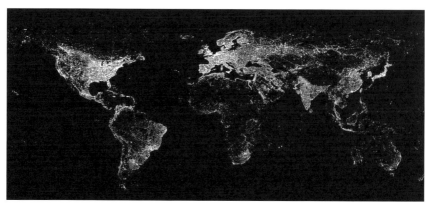

Figure 9.3 Earth night view from space with city lights over the whole Earth. NASA image

hundred thousand years compared with the millions of years that dinosaurs dominated life on land, the ability of humans to influence geological processes and cycles has become critical. The American astrophysicist Neil de Grasse Tyson said 'Dinosaurs are extinct today because they lacked opposable thumbs and the brainpower to build a space program'. The clear implication here is that dinosaurs would have avoided extinction if they had been smarter and possessed better manual dexterity. The dinosaurs dominated the Earth from the Triassic through the Jurassic and into the Cretaceous Period – a reign of almost 200 myr. This compares with a few tens of thousands of years at most that our species, *Homo sapiens*, has dominated the Earth. Since many scientists believe we are currently on a path leading to self-destruction it is perhaps a little early to be condescending about the intelligence level of dinosaurs.

The exact start of the new epoch, the Anthropocene, in Earth history and whether or not it is ratified by some official body is irrelevant. The fact is we are currently living in it. There is already sufficient evidence of that in the sand and mud of the seabed, in the river sediments, in the desert sands and in the layers of ice in the glaciers and ice caps to leave suitable marker horizons behind for future generations of geologists to interpret and perhaps wonder about.

Climate change: possible causes and implications

A feature of the Earth's atmosphere since the Industrial Revolution has been a steady increase in the amount of carbon dioxide (CO_2). The CO_2 content in the atmosphere has varied throughout geological time in response to a number of factors but the amount due to human activity – the anthropogenic contribution – is considered by many scientists to be producing changes which are larger and occurring at a faster rate than similar changes which occurred before the Industrial Revolution. Studies show that the levels of CO_2 in the atmosphere have risen steeply since the start of the Industrial Revolution in the late

18th century, with an accelerating rise in the late 20th century. This sharp rise in CO_2 levels in the atmosphere has triggered great debate in the early years of the 21st century, with phrases such as 'global warming', 'climate change' and 'carbon capture' becoming almost household terms. It has become clear that human activity has now reached a level that impinges on all aspects of the global ecosystem – the land areas, the oceans and the atmosphere. The changes in the atmosphere supply a convenient parameter for comparison between natural variation and that produced by anthropogenic activities.

From Figure 9.4 it can be seen that CO_2 levels have steadily risen since the late 19th century. At the same time as this increase in CO_2 there has been a gradual rise in the average global temperature – global warming in other words. The question is whether this global warming has a natural cause or whether it is anthropogenic or man-made due to the burning of fossil fuels or other factors. It is known that in the Middle Ages (from around the 12th century to the 15th century) there was what was known as the Medieval Warm Period when temperatures were probably higher than they are today. This period of warm climate was followed by the Little Ice Age, which lasted for about 300 years. The River Thames was famously frozen over in winter at this time, allowing fairs and other large gatherings to be held on the ice.

Possible natural causes for these significant variations in climate include cycles in the solar radiation pattern, heightened volcanic activity or changes in

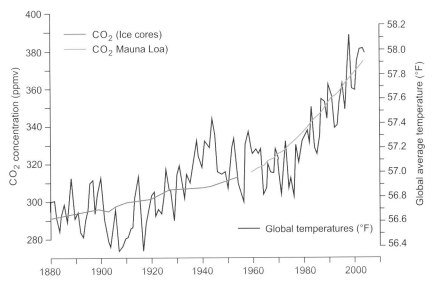

Figure 9.4 Global average temperature versus CO_2 concentrations, 1880–2004. CO_2 values in parts per million by volume (ppmv). After NASA.

ocean circulation. Detailed examination of the data shows a period from 1940 to 1970 when there was a distinct period of global cooling. This was despite continued rising levels of CO_2, so clearly other factors such as those noted above are involved.

One way of investigating the current extent of temperature change is by finding out what has happened in the past. Records of direct climatic measurement from thermometers, barometers or rain gauges have only been available since the mid-19th century. The climatic conditions that prevailed before then have to be reconstructed using indirect methods. These can include historical documents that describe the plant growing conditions at the time, unusual weather patterns or the extent and duration of ice cover for example. Recently developed techniques such as the examination of tree rings and ice cores can provide accurate indirect evidence of climate going back several thousands of years, as we have seen in the section on dendrochronology, but potentially much further back in the case of ice cores.

Lake Vostok is in East Antarctica, some 1,300 km from the South Pole (Fig. 9.5). It is the site of a project started in the early 1970s to extract ice cores as a way of studying ancient temperatures, CO_2 levels and volcanic dust and other features which may be of climatological interest. Lake Vostok is the largest of the sub-glacial lakes in Antarctica and lies under the surface of the central East Antarctica Ice Sheet at 3,488 m above sea level. The ice overlying the lake provides a palaeoclimatic record of 400,000 years based on examination of a 13-cm-diameter core drilled through 3,768 m and recovered by Russian scientists. The results of the drilling were published in 1995.

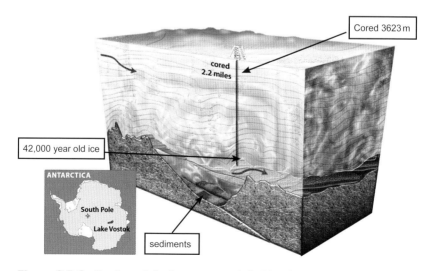

Figure 9.5 Section through the ice cover over Lake Vostok, Antarctica. USNSF image'.

In the upper part of the core individual layers can be counted, each representing a year. Deeper in the core the layers thin out due to pressure within the ice and individual layers are not recognizable. There are five peaks covering four cycles over the 400 kyr period with a cycle lasting around 100 kyr. The Vostok cores (Fig. 9.6) show CO_2 levels fluctuating from around 180 parts per million by volume (ppmv) to a maximum of 300 ppmv (blue line), corresponding to glacial and interglacial conditions. The red line shows temperature changes relative to a −56°C average from 1961 to 1990. From this it can be seen that the largest decreases in temperature from this line were of 8 to 9°C, while the largest increases were of 2 to 3°C.

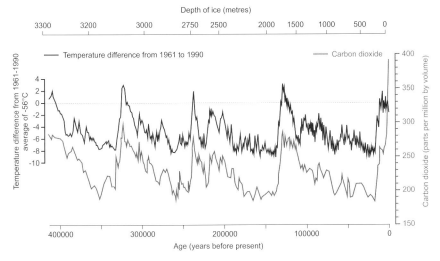

Figure 9.6 The Vostok ice core record. Carbon dioxide versus temperature for the last 420,000 years.

Detailed analysis of the graphs has generated controversy. Some climate researchers believe there is evidence that temperatures lead the CO_2 curves. In other words CO_2 is a climate follower. Other workers are convinced that higher temperatures follow on from increased CO_2 levels and therefore CO_2 is indeed a major cause of global warming.

Carbon dioxide comes from the burning of wood and fossil fuels such as coal, oil and natural gas. The rising levels of CO_2 are important because CO_2, along with methane (CH_4), nitrous oxide (N_2O), ozone (O_3) and water vapour are 'greenhouse gases'. These are so named because they form an insulating blanket over the Earth's surface which causes a higher proportion of the Sun's energy to be retained in the Earth's atmosphere. This results in a higher surface temperature than would occur with lower concentration of these gases.

Historical levels of CO_2 have fluctuated around 200–275 ppmv. Prior to the start of the Industrial Revolution and up to around the 1850s levels were relatively stable around 280 ppmv, but since then the concentration has risen to over 380 ppmv and is likely to reach 400 ppmv in the near future. In 2011 atmospheric CO_2 was 390 ppmv, nearly 100 ppmv above the previous maximum of around 300 ppmv recorded in the Vostok ice core about 323 kya, and the levels appear to be increasing by about 2 ppmv per year.

The ice core records show categorically that the rate of change of CO_2 levels is faster than at any other time in the last approximately 400 kyr and there is at least a possibility that these changes are anthropogenic. The debate is continuing; there are those who sense an approaching apocalypse as the Earth warms and the polar ice caps melt causing a worldwide rise in sea level while in the other camp are those who say the changes are not entirely anthropogenic and, while a reduction in fossil fuel use is desirable, the Earth is not yet headed for disaster. However, given the scale and rate of change over the last 200 years it is difficult to avoid the conclusion that the relationship between *Homo sapiens* and the world ecosystem is heading into unknown territory.

The collection of such a long ice core from Lake Vostok is a triumph of engineering and scientific endeavour and represents an invaluable data source. The fact that scientists are apparently capable of drawing directly opposite conclusions from that data means that they do not yet fully understand all the factors involved but that does not detract from their usefulness. We just have to be patient while the scientific debate runs its course, perhaps aided by additional information that becomes available with time.

The next few decades will be revealing in climatology. It is known that sunspot activity is currently low and if the Sun is the prime mover in climate change then this should mean we are entering at least a temporary period of global cooling. On the other hand if CO_2 is the principal influence on climate change, then the present trend of rising temperatures will continue. When the desire of countries such as China and India to attain the living standards regarded as the norm in North America and northern Europe is considered, along with the huge vested interests in the fossil fuel industries, it becomes apparent that this particular debate is a very long way from resolution. However, as noted in Chapter 3, the acquisition of consumer goods on a Western scale for the whole of an expanding world population is a physical impossibility. The scale of resources that would be required for this to happen, the raw materials and the energy supplies, would not be sustainable. The Great Acceleration of the second half of the 20th century, which saw a tripling of the world's population and the demand for natural resources soar, has placed the Earth's ecosystems under enormous strain and is an issue that must be addressed sooner rather than later. There should be a reduction in the emissions of greenhouse gases and a move away from our dependency on fossil fuels, of whatever type. We need to do this for all sorts of reasons, not least for reasons of political stability, for reasons of health and to try to mitigate the consequences of global warming on a scale that could involve more frequent extreme climate events and a rise in sea level which would have a devastating impact on many of the world's most densely populated areas.

The future of plate tectonics

The theory of plate tectonics shows us that we live on a dynamic planet which is constantly changing. Effects range from small-scale weathering and erosion processes to the movement of continents and the opening and closing of ocean basins. So while alterations in atmospheric composition, whether brought about by human activity or not, are potentially causing climate change with all the attendant ramifications, plate tectonic processes grind inexorably onwards.

Plate tectonics is still an active process, and will continue to reshape the face of the Earth in the future as it has done for much of the last 4 byr or so. It is possible to look forward in time to the consequences of some of these processes as currently recognized.

California will drift off

The earthquakes of the west coast region of the USA are the result of the frictional drag of the Pacific and North American Plates as they move past each other. The San Andreas Fault is responsible for many of the earthquakes in the San Francisco–Los Angeles area of California and forms a prominent topographical feature along its length (Fig. 9.7).

The map in Figure 9.8 shows the relative movement of the plates in this area. The main part of the North American plate is moving to the south-east (top left to bottom right) while the western portion is moving north-west (bottom right to top left). In time, if the directions of plate movement do not change, then Los Angeles will eventually become attached to San Francisco several million years from now.

The Mediterranean Sea will close

The European and African continents are parts of the world where current plate tectonic movements are likely to lead to interesting developments in the near future (geologically speaking that is). For a long time now the relative

Figure 9.7 Looking south-east along the surface trace of the San Andreas Fault in the Carrizo Plain, north of Wallace Creek. Elkhorn Road meets the fault near the top of the photo. USGS image.

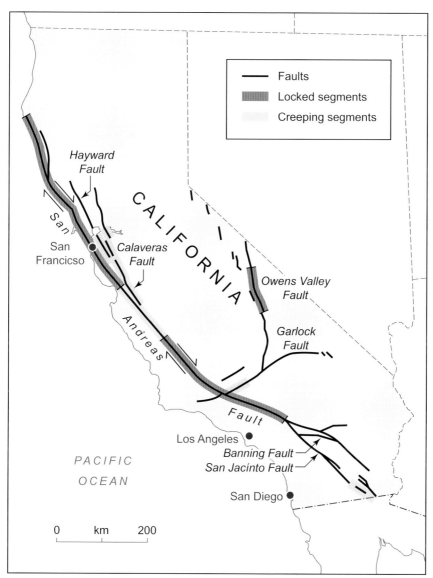

Figure 9.8 San Andreas Fault between San Francisco and Los Angeles, California. After USGS.

movement of Africa has been northwards, gradually reducing the extent of the Mediterranean Sea and forming linear mountain chains in southern Europe from Spain to Iran and including the Alps. This process is driven by spreading from the North Atlantic Ridge to the west, and the circum-Antarctic Ridge to the south, propelling Africa and Arabia northwards and gradually closing the ancient ocean of Tethys, one of the remnants of which is the Mediterranean Sea.

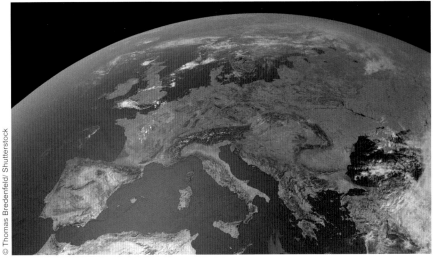

Figure 9.9 The Mediterranean Sea, a remnant of the ancient ocean of Tethys that existed between Laurasia and Gondwanaland around 200 million years ago.

Africa will separate from the Middle East

Further to the east, the interaction between Africa and Arabia is responsible for major crustal changes in East Africa, the Red Sea and the Gulf of Aden. With the East African Rift System, the Red Sea and the Gulf of Aden form the Afar triple junction, the meeting point of three tectonic plates (Fig. 9.10). The red arrows show the movement of the African, Arabian and Somalian Plates away from the Afar depression.

The Red Sea and the Gulf of Aden were formed as the Arabian Plate separated from part of the African Plate. The Gulf of Aden is aligned north-east–south-west near the triple junction, before curving round to east–west to link in with the mid-ocean ridge system in the Indian Ocean (see Fig. 9.10). The Red Sea runs roughly straight for 2,000 km northwards to where it joins the Aqaba–Dead Sea transform rift system.

The East African Rift System, which extends south-west from the triple junction in northeast Africa, is a developing divergent plate junction where the African Plate is splitting into the Nubian Plate and the Somalian Plate. The Arabian Plate is rotating in an anti-clockwise direction as the Red Sea widens by sea-floor spreading.

As a result of these plate movements the Red Sea is currently widening as Arabia is subducted northwards under Iran at the boundary with the Eurasian Plate, at a rate of a few centimetres per year. This collision formed the Zagros mountain belt, with its tightly folded sediments caused by the impact of Arabia and Asia meeting (Fig. 9.11).

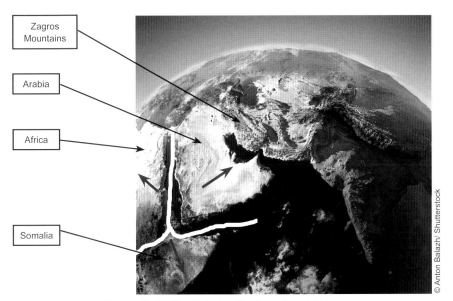

Zagros Mountains

Arabia

Africa

Somalia

© Anton Balazh/ Shutterstock

Figure 9.10 The triple junction in northeast Africa.

Figure 9.11 Tightly folded sediments in the Zagros Mountains, Iran. NASA image.

There are few better regions in the world than this to illustrate the collective impact of the dynamism of plate tectonics. From the eastern Mediterranean, south to East Africa and the Red Sea, south-east to Arabia and north to Iran there are examples of continental rifting, sea-floor spreading and an active subduction zone.

Extrapolating from these processes

By extrapolation of these processes we can make an educated guess at the likely distribution of continents and oceans in 50 myr from now. If we continue present-day plate motions the Atlantic will widen, Africa will collide with Europe closing the Mediterranean, Australia will collide with South East Asia, and California will slide northward up the coast to Alaska.

However, predicting the future distribution of continents on the Earth's surface is a bit like forecasting the weather – a 5-day forecast can be quite accurate but the forecast for the next month is an entirely different proposition. To predict the arrangement of continents beyond the next 50 myr is fraught with difficulties due to the increasing number of possible scenarios as the tectonic plates move in response to the internal forces coming from the core and the mantle. Once again we need to look back before looking forward and to examine the largest scale of time's cycle, the supercontinent cycle.

The next supercontinent

The term supercontinent was first mentioned in Chapter 1, when examining the complex geological history of the north-west of Scotland. That particular supercontinent was Rodinia, thought to have existed around 1.1 bya. It was probably mostly in the southern hemisphere and began to break up some time after 750 mya. This break-up led to the formation of the Iapetus Ocean about 600 mya and the opening and closing of this ocean over a period of about 200 myr, part of the Wilson Cycle. In this cycle, continents split to form new oceans which reach a maximum width before closing again in a continental collision zone that produces a mountain belt. This then forms a stable continent which splits to restart the cycle. This continent/ocean cycle, which operates on a timescale of hundreds of millions of years, is part of the supercontinent cycle, the periodic amalgamation and then break-up of the Earth's continents.

There were likely to have been supercontinents before Rodinia. For example, there is some evidence for a supercontinent named Ur which existed around 3 bya, but the further back in time we look the evidence becomes less and less conclusive. More convincing evidence exists for a supercontinent Nuna around 1.8–1.5 bya. The dispersal of the continents after the break-up of Rodinia led to the formation about 270 mya of the supercontinent Pangaea, about which we have most information and whose break-up about 200 mya gave us the distribution of continents we are familiar with today.

A timeline for the supercontinent cycle could be drawn up as follows:

Next supercontinent +250 myr

_ Present
_ Pangaea 270 mya

_ Rodinia 1.1 bya
_ Nuna ~1.8 bya

_ Ur ~3 bya

On the evidence of the construction and destruction of previous supercontinents therefore we are currently about half-way through the supercontinent cycle. We can expect the continents to move together over the next 250 myr. From what geologists have learned about past supercontinents what can they say with any certainty about the next one and what is its likely configuration?

The projection for 50 myr ahead (Fig. 9.12A) is mostly a continuation of the present processes, with one notable exception. It has been assumed that by +50 myr from now the Atlantic will have reached its maximum width and started to contract via a subduction zone situated off the east coast of North America, with an associated western Atlantic Trench. While this is a perfectly plausible development it is by no means certain. If we extend the time in future to 150 myr ahead the result is a narrowing of the Atlantic Ocean with the African continent moving northwards and Australia joined with Antarctica and moving northwards towards Asia (Fig. 9.12B).

This sequence represents one possible outcome and would lead to a new supercontinent, not unlike Pangaea, by 250 myr ahead (Fig. 9.12C). The sequence of events leading to such an outcome has been described as introversion, where the Atlantic closes and the new supercontinent forms by the continents rolling back in a near-reversal of their tracks during the dispersion. This projected new supercontinent has been variously named as Neopangaea, Pangaea Ultima or Pangaea II.

An alternative version of the build-up to the next supercontinent is to assume the Atlantic Ocean does not contract, but in fact keeps on expanding. This is as plausible as the alternative. The prediction of the development of new subduction zones is very difficult – the forces involved are complex and include factors such as existing crustal weaknesses, sediment loading on the sea floor and the rate of spreading from the ridge. If the Atlantic continues to expand, then the Pacific shrinks and the Americas will collide with Asia. This version of events has been labelled extroversion and means that the coalescence of the continents would take place on the far side of the planet from their original positions.

Consequences and implications of future changes

Whether by introversion or extroversion around 250 myr from now a visitor watching from space would see a single supercontinent on the face of the Earth surrounded by a globe-encircling ocean. In what ways will this new environment differ from the one we are familiar with today? In general global sea levels are lower when continents are together and higher when they are dispersed. This is because active spreading ridges take up more space when heated, cooling and shrinking with time as the crust moves away from the ridge crest.

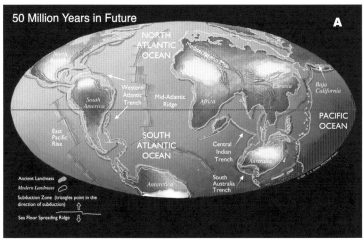

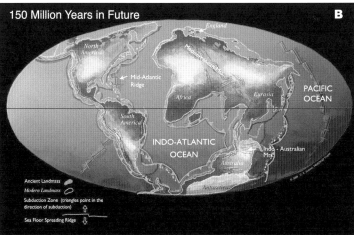

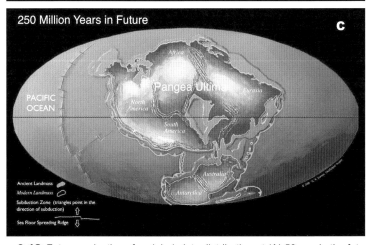

Figure 9.12 Future projections for global plate distribution at (A) 50 myr in the future, (B) 150 myr in the future and (C) 250 myr in the future. C.R. Scotese, PALEOMAP Project.

If the sea floor is active and young it will be relatively shallow and sea level will therefore be high. If, however, the sea floor is generally old and lacking in active spreading ridges then the ocean bed will be relatively deep and overall sea level will be low, exposing more of the continents.

The configuration of and distribution of the continents on the surface of the Earth has a large influence on climate, affecting the global wind patterns and the speed and direction of ocean currents. In addition land areas have a higher albedo than oceans. They reflect more of the Sun's energy causing changes in coastal wind patterns. Because the centres of large continents are generally higher in elevation than their coastal regions they tend to be cooler and drier. This was the case for the middle of Pangaea and the same can be seen today in the vast, cold Gobi desert in Mongolia and China. In general supercontinents tend to create extremes of environment with a tendency to develop violently changing weather patterns. There is a likelihood of monsoonal storms because of the temperature differences between the large land and sea areas.

There are a number of changes that are currently taking place where it is reasonable to extrapolate into the future, even a future as far ahead as the next supercontinent. It was noted earlier that the rotation of the Earth is slowing down, mainly due to friction in the oceans and so the length of the day is steadily shortening. If the Devonian day was several hours shorter than now 400 mya, then this trend will continue and at the time of the next supercontinent a day will be several hours shorter than the current 24.

The use of fluctuations in the geomagnetic field as a tool in magnetochronology was examined in Chapter 8. The Geomagnetic Polarity Time Scale (GPTS) shows that the last major reversal in the polarity of the geomagnetic field was some 780 kya and this is one of the longest periods without a reversal in the past 5 myr or so. In the last 10 myr reversals have occurred on average four or five times per million years. This means that going on past evidence a reversal of the geomagnetic field is overdue. It can be assumed that by the time of the next supercontinent there will have been a number of reversals. Since there appears to be no evidence from the fossil record of serious harm to species, including our own, from previous reversals, then it is probable that reversals will not be a force for genetic mutation in the future.

Potential impact events in the future

Many impact events, instances of objects from space colliding with the surface of the Earth, have significantly shaped the planet's history. One such event is likely to have been responsible for the formation of the Earth–Moon system and the impact of the Chicxulub asteroid in Central America 66 mya was certainly a factor in the mass extinction at the end of the Cretaceous Period. There

Figure 9.13 Close-up view of Eros, an asteroid with a near-Earth orbit. NASA image.

are still many large asteroids in the Solar System, such as Eros, an asteroid with a near-Earth orbit (Fig. 9.13).

Objects between 1 and 2 km in diameter are considered to be on the critical threshold for a global catastrophe. The impact of objects this size and above will cause total destruction at the impact site (the Chicxulub crater is 180 km wide and 10 km deep) but will also release so much gas and dust into the atmosphere that sunlight and therefore plant growth would be seriously affected globally. This would lead to worldwide famine and the likely extinction of large land animals. An impact in the ocean rather than on land would have devastating effects on marine life and the generation of tsunamis from the impact would inundate coastal areas around the world.

Figure 9.14 shows the relationship between the size of an Earth-asteroid impact and the frequency of such an event. The lower threshold for a global catastrophe is estimated to be an object of around 2 km diameter and that such an event could be expected to occur every 1 to 10 myr. In the next 250 myr before the formation of the next supercontinent it seems likely, again given past history, that the Earth will suffer an impact event with a celestial body greater than 2 km diameter, with all the consequences for the environment that implies.

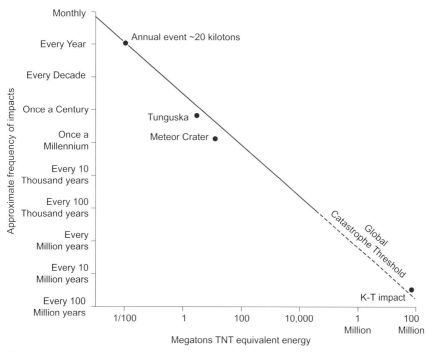

Figure 9.14 Graph showing the relationship between the size of an Earth-asteroid impact and the frequency of such an event. After NASA.

The future for *Homo sapiens*

The elephant in the room in all this speculation, the topic that so far has not been addressed, is whether or not any members of *Homo sapiens* will be present on this supercontinent and will there even be any signs of the previous existence of the species. In the light of the figures quoted above, and leaving aside the species' ability to cause major environmental damage, to engage in mutually destructive warfare and to engender global pandemics, it would take a great optimist to predict a significant human presence on the newly created supercontinent 250 myr from now. When we consider how long our species has been around in the full context of geological time we realize that, despite our capacity to control and influence our environment like no other species before us, we still represent such a miniscule proportion of geological time as to be insignificant.

However, the recent developments in genetic engineering and nanotechnology mean that for the first time in Earth history the laws of natural selection are being modified, if not broken. Humans now have the ability to modify the capabilities and other physical attributes of an organism by gene implantation, for example, in order to satisfy some perceived need or notion of change. Although there currently appears to be a reluctance to experiment on people it can only be a matter of time before such biological engineering moves from gene modification to cure medical conditions such as dementia or Parkinson's Disease to a more comprehensive alteration of the characteristics of *Homo sapiens*. It is not inconceivable that such modifications could effectively produce a new species or subspecies. As has often been the case over the last 100 years or so, our technological capability is out-stripping our ability to make considered and clear-sighted use of the opportunities becoming available. Perhaps then it will not be *Homo sapiens* living on the next supercontinent but a descendant species, in much the same way as *Homo sapiens* evolved from earlier humanoid forms, all of whom are now extinct.

Epilogue

Currently we are greatly concerned with the future of our species. There are various dire predictions as to the impact on future generations of our consumption of non-renewable resources, of the effects of burning fossil fuels and of the implications of a burgeoning world population in terms of food and water supply. In general these worries refer to the next few tens or at most hundreds of years. As a species we tend to be aware of five generations – two behind us, two ahead of us with an existentialist emphasis on the middle generation we are part of. In reality most people know little of their ancestors further back than their grandparents and cannot project further ahead than their grandchildren.

Yet what we have been examining so far about deep time should inform us that the timescale of our disquiet is miniscule. We arrived in the last minute before midnight on 31 December, or at the end of the fingernail of the out-stretched arm, or the outermost limb of the spiral of time – depending on which metaphor we are using to illustrate the vastness of geological time.

If we really want to look ahead on the same scale as we have looked back into deep time, then we need to be much more realistic. Compared with the effects of an eruption of the supervolcano under Yellowstone National Park in the USA, or the impact of even a smallish asteroid, the changes in climate and sea level we are currently agonizing over are really minor inconveniences.

Figure 9.14 showed that an asteroid impact event is overdue. Given the likely planet-wide effects of such an impact even a worldwide nuclear war would seem relatively minor. A natural catastrophe from a supervolcano eruption or an asteroid impact could occur at almost any time in the next few hundreds or thousands of years.

Devastating though these events will be – I say *will* because they have happened before and their recurrence is inevitable – they are unlikely to affect the processes of plate tectonics. These will continue to rearrange the distribution of the Earth's continents and oceans for as long as the internal heat engine still has energy. In so far as the past points to the future there will be a new supercontinent formed around 250 myr from now. Having all the Earth's land masses together has major implications for sea levels, the pattern of ocean circulation, weather patterns and terrestrial environments.

Whatever the conditions on the next version of Pangaea it is unlikely that *Homo sapiens* will be around to experience them. We will have joined the long list of species to originate, develop and then expire during the course of geological history. Some have been more successful in surviving than others and some have made a greater impact on the Earth's environment than others, but in the end most have succumbed to the inevitable and become extinct. If we are looking to past events in order to predict the future, it is difficult to see how *Homo sapiens* could be capable of surviving the events of both time's arrow and time's cycle. The one proviso is that *Homo sapiens* changes markedly to adapt to changing conditions or even leaves planet Earth entirely.

In attempting to predict the future of the planet even further ahead than the next supercontinent, we begin to move into the field of astronomy. Remember, astronomers operate on even longer timescales than geologists. Their prediction for Earth, based on events elsewhere in the universe, is that as the Sun runs out of fuel in a few billion years it will expand and turn into a red giant. Life on Earth will become impossible because of the greatly increased outflow of energy. If we leave aside the future as envisaged by the genetic engineers and the science fiction writers our only alternative is *carpe diem* – seize the day – and use the time available to us on the planet as productively as possible.

I have endeavoured in this book to introduce the reader to the concepts of geological time and Earth history and how this narrative can be read from the landscape. John McPhee in *Basin and Range* mentioned how geologists talk about seeing the 'picture' or sometimes the 'big picture'. This 'picture' is the sum of things like the plate motions, the rock types, the fossil assemblages and the palaeo-environments. The problem is that most of the picture is missing and geologists have had to become adept at dealing with incomplete data sets. If I have had any success in my aspiration to explain time and landscapes, if at least some readers have an increased awareness of the extent of deep time and how important it is to our view of the world we live in, then I have some concern that such enhanced awareness may come at a price. Once the available evidence has been evaluated and the big picture visualized, the relatively insignificant place of *Homo sapiens* as a species in this grand scheme becomes apparent. This can present a challenge for many people. The realization that we represent such a small episode in the overall history of the Earth, albeit a species with an increasingly influential role, can be difficult to reconcile with an individual's view of themselves and their purpose in life. This particularly applies if that view of self is combined with a religious belief that places humans at the centre of affairs.

Still, there is much to be said for knowing one's place, especially when that place is part of a continuum that has lasted 4.5 byr and is ongoing. The Earth has passed through a remarkable sequence of events and processes during that time and as a species we have undergone a remarkable series of revolutions. The Cognitive Revolution at 70 kya set us apart from other primates and began the fascination with the measurement of time. The Agricultural Revolution about 12 kya, set us on the path to modern life styles and society found it had time on its hands to continue the exploration of time and its significance. The Scientific Revolution, starting around 500 years ago, began to challenge the ideas of the early Greek and Roman philosophers and the orthodoxies of the religions of the time, and laid the foundations of modern scientific thought. The Industrial Revolution, starting in Britain in the late 18th century, opened up a whole new world order, fuelled by new energy sources that built an industry-based society that continues to develop today. Technological and scientific progress has continued without break since that time and the so-called Great Acceleration of the last 50 years or so only serves to underline this.

Our place, therefore, is at the forefront of this revolution, for the present time at least. What awaits us in the future we can only speculate on, but with the concepts of time's arrow and time's cycle we can at least make an educated guess.

References and further reading

Albritton, C. C. (1980) *The Abyss of Time: Changing Conceptions of the Earth's Antiquity After the Sixteenth Century.* San Francisco: Freeman, Cooper and Company

Allen, J. E. and Burns, M. (1986) *Cataclysms on the Columbia.* Timber Press, Inc: Portland, Oregon

Atwater, B. F. (1986) Pleistocene glacial-lake deposits of the Sunpoil River Valley, northeastern Washington. *US Geological Survey Bulletin* **1661**

Baillie, M. G. L. (1999) *Exodus to Arthur: Catastrophic Encounters With Comets.* London: B. T. Batsford

Bretz, J H. (1923) The Channeled Scabland of the Columbia Plateau. *Geology* **31**, 617–49

Frank, A. (2012) *About Time: From Sun Dials to Quantum Clocks, How the Cosmos Shapes Our Lives – and We Shape the Cosmos.* Oxford: Oneworld Publications

Gallup (2010) www.gallup.com/poll/145286/four-americans-believe-strict-creationism.aspx (accessed April 2015)

Gould, S. J. (1987) *Time's Arrow, Time's Cycle: Myth and Metaphor in the Discovery of Geological Time.* Cambridge: Harvard University Press

Harari, N. H. (2014) *Sapiens: A Brief History of Humankind.* London: Harvill Secker

Heaney, S. (2004) Bog bank, rock face and the far fetch of poetry. In Parkes, M. A. (ed) (2004) *Natural and Cultural Landscapes – The Geological Foundation*, Dublin: Royal Irish Academy, pp. 11–17

Holland, C. H. (1999) *The Idea of Time.* Chichester: Wiley

Hunt, C. B. (1959) Dating of mining camps with tin cans and bottles. *Geotimes* **3** (8), 8–10

Libarkin, J., Kurdziel, J. P. and Anderson, S. W. (2007) College student conceptions of geological time and the disconnect between ordering and scale. *Journal of Geoscience Education* **55**, 413–22

McElhinny, M. W. and Senanayake, W. E. (1982) Variations in the geomagnetic dipole 1, the past 50,000 years. *Journal of Geomagnetism and Geoelectricity* **34**, 39–51

McPhee, J. (1981) *Basin and Range.* New York: Farrar, Strauss and Giroux

Nield, T. (2007) *Supercontinent: Ten Billion Years in the Life of Our Planet.* London: Granta Books

Playfair, J. (1964) *Illustrations of the Huttonian Theory of the Earth: A Facsimile Reprint.* New York: Dover

Rudwick, M. (2005) *Bursting the Limits of Time: The Reconstruction of Geohistory in the Age of Revolution.* Chicago: University of Chicago Press

Rudwick, M. (2008) *Worlds Before Adam: The Reconstruction of Geohistory in the Age of Reform.* Chicago: University of Chicago Press

Stuiver, M. and Pearson, G. W. (1993) High-precision bidecadal calibration of the radio-carbon time scale, AD 1950–500 BC and 2500–6000 BC. *Radiocarbon* **35** (1), 1–23

Winchester, S. (2001) *The Map that Changed the World: A Tale of Rocks, Ruin and Redemption.* London: Viking

Wyse Jackson, P. N. (2006) *The Chronologer's Quest: Episodes in the Search for the Age of the Earth.* Cambridge: Cambridge University Press

Index

Note: page numbers in *italic* denote illustrations.

Index

Index

Index

Illustration credits

Illustration credit lines appear by the illustration. This list gives further details as appropriate. All other illustrations in this book are © the author.

Half title page verso: Ptolemaic orbits. Plate 3 from Harmonia Macrocosmica, Andreas Cellarius, 1661. Source www.staff.science. uu.nl/~gent0113/cellarius/cellarius_plates.htm

2.2 Visualization of deep time. From the Boulder Community Network, Visualizing Deep Time. http://bcn.boulder.co.us/basin/local/sustain2a.html

2.4 Geological spiral time line. See https://commons.wikimedia.org/wiki/File:Geologica_time_USGS.png (courtesy United States Geological Survey)

2.6 Aerial view of the Nile Delta. Jacques Descloitres, MODIS Rapid Response Team, courtesy NASA/GSFC

3.1 Bingham Canyon Mine, Utah. Photograph courtesy of Earth Science and Remote Sensing Unit, NASA Johnson Space. http://eol.jsc.nasa.gov/SearchPhotos/photo.pl?mission=ISS015&roll=E&frame=29867

3.2 Dust Bowl conditions, South Dakota. Sloan (?) – courtesy United States Department of Agriculture; Image Number: 00di0971

3.4 Artesian aquifer. after Andrew Dunn. https://commons.wikimedia.org/wiki/File:Artesian_Well.svg

4.1 Ptolemy's model of the universe. Attribution : By Bartolomeu Velho (own work) [Public domain}, via Wikimedia Commons

4.2 Portrait of Galileo. Attribution : Justus Sustermans [Public Domain], via Wikimedia Commons

4.3 Portrait of Rene Descartes. Attribution : After Frans Hals (1582/3-1666) [Public Domain], via Wikimedia Commons

4.4 Portrait of Johannes Kepler. Attribution : Unknown [Public Domain] via Wikimedia Commons

4. 6 Portrait of James Ussher, Archbishop of Armagh after Peter Lely © **National Portrait Gallery, London** NPG 574

4.7 First page of the Annales Veteris Testamenti. Attribution : By James Ussher (Wing, Early English Books 1641-1700) [Public Domain], via Wikimedia Commons

4.8 Title page of Telluris Theoria Sacra. Attribution: Thomas Burnet [CC BY 4.0 (http://creativecommons.org/licenses/by/4.0)], via Wikimedia Commons

5.1 The Himalayas and Tibetan Plateau viewed from space. NASA image courtesy Jeff Schmaltz, MODIS Rapid Response Team, Goddard Space Flight Center.

5.2 Distribution of tectonic plates on the Earth's surface. Adapted from Park, Introducing Tectonics, Rock Structures and Mountain Belts, 2012, Dunedin Academic Press, Edinburgh.

5.3 Types of plate boundaries. After United States Geological Survey illustration by Jose F. Vigil (http://pubs.usgs.gov/gip/dynamic/Vigil.html)

5.4 Convection cells in the Earth's mantle. Adapted from the United States Geological Survey illustration at http://pubs.usgs.gov/gip/dynamic/unanswered.html

5.5 Formation of the Himalayas by collision of India with Asia. After United States Geological Survey illustration at http://pubs.usgs.gov/gip/dynamic/himalaya.html

5.7 Stages of the Wilson Cycle. After Hannes Grobe. http://www.slideshare.net/zombraweb/wilson-cycle

5.8 Breakup of Pangaea. After United States Geological Survey illustration. http://pubs.usgs.gov/gip/dynamic/historical.html

5.10 Grand Canyon from space. Photograph courtesy of ISS Crew Earth Observations Facility and the Earth Science and Remote Sensing Unit, Johnson Space Center.

5.11 First camp of the second John Wesley Powell expedition. Photograph by E. 0. Beaman [Public domain or Public domain], via Wikimedia Commons

5.13 Geological succession at the Grand Canyon. After United States Geological Survey, http://education.usgs.gov/images/schoolyard/GrandCanyonAge.jpg

5.16 Impact craters on the moon. Photograph courtesy of NASA. Credit : NASA/ESA/ D.Ehrenreich

5.18 Site of Chicxulub impact crater, Gulf of Mexico. After NASA Astrophysics Science Division, Goddard Space Flight Center

5.19 Trees felled at Tunguska, 1908. Photograph courtesy of NASA.

5.20 Left. Charles Lyell by George J. Stodart

5.20 Right. Georges Cuvier. Attribution: By Georges_Cuvier.jpg: Unknown derivative work: Beao (Georges_Cuvier.jpg) [Public domain], via Wikimedia Commons

6.1 Steno's four laws stratigraphy. After Jones, Introducing Sedimentology, 2015, Dunedin Academic Press, Edinburgh.

6.3 Range chart for variation of nails, cans and bottles in Californian mining camps. After Hunt, C.B. 1959 Dating of mining camps with tin cans and bottles. Geotimes, 3, 8

6.4 Steno's drawing of a shark's head and a so-called tongue stone. Courtesy of NASA : earthobservatory.nasa.gov

6.5 Robert Hooke's microscope. Billings Microscope Collection, National Museum of Health and Medicine, Armed Forces Institute of Pathology.

Illustration credits

6.6 Portraits of Alexander von Humboldt. Main picture: Portrait by Friedrich Georg Weitsch, 1806. Attribution: Friedrich Georg Weitsch [Public domain], via Wikimedia Commons

6.8 First geological map of Great Britain, published by William Smith, 1815. From http://www.livescience.com/449-map-changed-world.html

6.9 Stratigraphic ranges and origins of some major animal and plant groups. After pubs.usgs.gov/gip/fossils/fig11.gif

6.12 Geological column showing the main subdivisions and periods of geological time. After Wyse Jackson, Introducing Palaeontology, 2010, Dunedin Academic Press, Edinburgh

6.15A Twistleton Scar glacial erratic, Yorkshire Dales National Park. © Phil MacD Photography/ Shutterstock

6.15B Lateral moraines above Lake Louise, Alberta, Canada. Photograph by Mark A. Wilson, The College of Wooster

7.1 The Great Nebula in Orion. Photograph courtesy of NASA, Astronomy Picture of the Day, Stefan Seip.

7.2 Cartoon illustrating the origin of presolar grains within a planetary nebula. After Manavi Jodhar, 2009, PhD thesis Washington University in St Louis. Original diagram by Larry Nittler.

7.3 William Thomson, Baron Kelvin of Largs. Attribution: Author unknown [Public domain], via Wikimedia Commons

7.4 Wilhelm Conrad Röntgen. Attribution: Author unknown [Public domain], via Wikimedia Commons

7.5 One of the first human X-ray photographs, 1896. Attribution: By Wilhelm Röntgen; originally uploaded to en.wikipedia by E rulez. (Transferred from en.wikipedia.) [Public domain], via Wikimedia Commons

7.9 Ernest Rutherford. George Grantham Bain Collection (Library of Congress). LC-DIG-ggbain-36570.

7.13 Calibrating the stratigraphical column using radiometric dates. After illustration in Earth History 1, Science Foundation Course, the Open University, 1971.

7.14 Stratigraphical column: the left hand section shows the true extent of Precambrian time. After Upton, *Volcanoes and the Making of Scotland*, 2015, Dunedin, Edinburgh

7.15 from Introducing Palaeontology, Patrick Wyse Jackson, 2010, Dunedin, Edinburgh

8.1 The eruption of Mount Pinatubo in the Philippines. United States Geological Survey see http://pubs.usgs.gov/fs/1997/fs113-97/.

8.3 Distribution of the ash cloud over Europe a few days after the eruption of Eyjafjallajökull in 2010. Upper photograph courtesy of NASA/NOAA. Map after BBC News, source of information the Meteorological Office.

8.8 Tephra horizons, from Thor Thordarson & Ármann Höskuldsson, *Iceland 2nd Edition*, 2014, Dunedin, Edinburgh

8.11 Scanning electron microscope photograph of growth rings in black coral from the Gulf of Mexico. United States Geological Survey. http://gallery.usgs.gov/photos/03_28_2011_fja4Dpo0CW_03_28_2011_2

8.12 Fossilized horn coral (Heliophyllum) of Devonian age, showing growth rings. United States Geological Survey. http://gallery.usgs.gov/images/04_30_2014/xc61WjiVUp_04_30_2014/medium/Animals_Cnidarians_00005.jpg

8.17 Plots of true age (from tree rings) against radiocarbon age. After Stuiver, M. and Pearson, G.W., 1993, Radiocarbon, 35, 1, pp 215-230.

8.22 The Earth's solid inner core and the outer core with the helical convection cells which create circulating electrical currents which generate the magnetic field. After United States Geological Survey image.

8.23 Earth's Phanerozoic magnetic polarity timescale. After Vita-Finzi & Fortes, Planetary Geology, 2013, Dunedin Academic Press, Edinburgh

8.24 A theoretical model of the formation of magnetic striping. After an image from the United States Geological Survey, http://pubs.usgs.gov/gip/dynamic/developing.html

8.25 The geomagnetic Polarity Time Scale for the last 5 myr. After United States Geological Survey Open-File Report 03-187

8.27 Magnetic north pole positions of the Earth for the recent and historical past. After Vita-Finzi & Fortes, Planetary Geology, 2013, Dunedin Academic Press, Edinburgh

8.28 Variations in the strength of the dipole field over the last 10 kyr for archaeological hearth in Europe. After McElhinny, M.W. and Senanayake, W.E. 1982, Journal of Geomagnetism and Geoelectricity, 34, 39-51

8.29 The Laetoli footprints, left by early hominids on a layer of freshly deposited ash in Tanzania, nearly 4mya: replica of footprint trail, Smithsonian's Human Origins Program

Figure 9.1 Earth night view from space with city lights over Europe and the Middle East. Photograph courtesy of NASA's Earth Observatory: http://eoimages.gsfc.nasa.gov/images/imagerecords/55000/55167/earth_lights.gif

9.3 Earth night view from space with city lights over the whole Earth. Photograph courtesy of NASA's Earth Observatory

9.4 Global average temperature versus CO_2 concentrations, 1880-2004. Adapted from NASA image: https://oco.jpl.nasa.gov/images/ocov2/CO2-Temp.jpg

9.5 (Section through the ice cover over Lake Vostok, Antarctica. By Nicolle Rager-Fuller / United States National Science Foundation

9.6 The Vostok ice core record. Carbon dioxide versus temperature for the last 400kyr. Based on data from the United States National Ice Core Laboratory.

9.7 Looking south-east along the surface trace of the San Andreas Fault in the Carrizo Plain, California. Image courtesy of the United States Geological Survey.

9.8 San Andreas Fault between San Francisco and Los Angeles, California. Adapted from a map produced by the United States Geological Survey, pubs.usgs.gov

9.11 Tightly folded sediments in the Zagros Mountains, Iran. Photograph courtesy of NASA, Goddard Space Flight Center: http://disc.sci.gsfc.nasa.gov/geomorphology/GEO_2/geo_images_T-42/FigT-42.7.jpeg

9.13 Close-up view of Eros, an asteroid with a near-Earth orbit. Photograph from NASA 's Near Earth Program: https://solarsystem.nasa.gov/planets/profile.cfm?Object=Asteroids

9.14 Graph showing the relationship between the size of an Earth-asteroid impact and the frequency of such an event. After a NASA graph: https://solarsystem.nasa.gov/scitech/display.cfm?ST_ID=345